NATIONAL ACADEMIES *Sciences Engineering Medicine*

NATIONAL ACADEMIES PRESS
Washington, DC

Advanced Battle Management System

Needs, Progress, Challenges, and Opportunities Facing the Department of the Air Force

Ellen Y. Chou, *Editor*

Committee on Air Force Advanced Battle Management System

Air Force Studies Board

Division on Engineering and Physical Sciences

Consensus Study Report

THE NATIONAL ACADEMIES PRESS   500 Fifth Street, NW   Washington, DC 20001

This activity was supported by Contract No. FA9550-16-D-0001/FA8650-20-F-9314 with the U.S. Air Force. Any opinions, findings, conclusions, or recommendations expressed in this publication do not necessarily reflect the views of any organization or agency that provided support for the project.

International Standard Book Number-13: 978-0-309-68621-1
International Standard Book Number-10: 0-309-68621-0
Digital Object Identifier: https://doi.org/10.17226/26525

This publication is available from the National Academies Press, 500 Fifth Street, NW, Keck 360, Washington, DC 20001; (800) 624-6242 or (202) 334-3313; http://www.nap.edu.

Copyright 2022 by the National Academy of Sciences. National Academies of Sciences, Engineering, and Medicine and National Academies Press and the graphical logos for each are all trademarks of the National Academy of Sciences. All rights reserved.

Printed in the United States of America.

Suggested citation: National Academies of Sciences, Engineering, and Medicine. 2022. *Advanced Battle Management System: Needs, Progress, Challenges, and Opportunities Facing the Department of the Air Force*. Washington, DC: The National Academies Press. https://doi.org/10.17226/26525.

The **National Academy of Sciences** was established in 1863 by an Act of Congress, signed by President Lincoln, as a private, nongovernmental institution to advise the nation on issues related to science and technology. Members are elected by their peers for outstanding contributions to research. Dr. Marcia McNutt is president.

The **National Academy of Engineering** was established in 1964 under the charter of the National Academy of Sciences to bring the practices of engineering to advising the nation. Members are elected by their peers for extraordinary contributions to engineering. Dr. John L. Anderson is president.

The **National Academy of Medicine** (formerly the Institute of Medicine) was established in 1970 under the charter of the National Academy of Sciences to advise the nation on medical and health issues. Members are elected by their peers for distinguished contributions to medicine and health. Dr. Victor J. Dzau is president.

The three Academies work together as the **National Academies of Sciences, Engineering, and Medicine** to provide independent, objective analysis and advice to the nation and conduct other activities to solve complex problems and inform public policy decisions. The National Academies also encourage education and research, recognize outstanding contributions to knowledge, and increase public understanding in matters of science, engineering, and medicine.

Learn more about the National Academies of Sciences, Engineering, and Medicine at **www.nationalacademies.org**.

Consensus Study Reports published by the National Academies of Sciences, Engineering, and Medicine document the evidence-based consensus on the study's statement of task by an authoring committee of experts. Reports typically include findings, conclusions, and recommendations based on information gathered by the committee and the committee's deliberations. Each report has been subjected to a rigorous and independent peer-review process and it represents the position of the National Academies on the statement of task.

Proceedings published by the National Academies of Sciences, Engineering, and Medicine chronicle the presentations and discussions at a workshop, symposium, or other event convened by the National Academies. The statements and opinions contained in proceedings are those of the participants and are not endorsed by other participants, the planning committee, or the National Academies.

Rapid Expert Consultations published by the National Academies of Sciences, Engineering, and Medicine are authored by subject-matter experts on narrowly focused topics that can be supported by a body of evidence. The discussions contained in rapid expert consultations are considered those of the authors and do not contain policy recommendations. Rapid expert consultations are reviewed by the institution before release.

For information about other products and activities of the National Academies, please visit www.nationalacademies.org/about/whatwedo.

COMMITTEE ON AIR FORCE ADVANCED BATTLE MANAGEMENT SYSTEM

PHILIP S. ANTÓN, Stevens Institute of Technology, *Chair*
SHARON A. BEERMANN-CURTIN, STRATCON, LLC
MICHAEL A. FANTINI, U.S. Air Force (retired)
PRISCILLA E. GUTHRIE, Institute for Defense Analyses
PAUL G. KAMINSKI, NAE,[1] Technovation, Inc.
THOMAS A. LONGSTAFF, Carnegie Mellon University
KATHARINA G. McFARLAND, Blue Oryx, Inc.
GUNASEKARAN SEETHARAMAN, Naval Research Laboratory
DAVID M. VAN BUREN, Crossroads Management, LLC

Staff

ELLEN Y. CHOU, Board Director, *Study Director*
EVAN ELWELL, Research Associate
RYAN MURPHY, Program Officer

See Appendix E, Disclosure of Unavoidable Conflicts of Interest.

---

[1] Member, National Academy of Engineering.

AIR FORCE STUDIES BOARD

ELLEN M. PAWLIKOWSKI, NAE,[1] Independent Consultant, *Chair*
KEVIN G. BOWCUTT, NAE, Boeing Company
CLAUDE CANIZARES, NAS,[2] Massachusetts Institute of Technology
MARK F. COSTELLO, Georgia Institute of Technology
WESLEY L. HARRIS, NAE, Massachusetts Institute of Technology
JAMES E. HUBBARD, JR., NAE, Texas A&M University
LESTER L. LYLES, NAE, U.S. Air Force (retired)
WENDY M. MASIELLO, Wendy Mas Consulting, LLC
LESLIE A. MOMODA, HRL Laboratorios, LLC
OZDEN OCHOA, Texas A&M University
F. WHITTEN PETERS, Williams and Connolly, LLP
HENDRICK RUCK, Edaptive Computing, Inc.
JULIE J.C.H. RYAN, Wyndrose Technical Group
MICHAEL SCHNEIDER, Lawrence Livermore National Laboratory
GRANT STOKES, NAE, Massachusetts Institute of Technology

Staff

ELLEN Y. CHOU, Director
GEORGE COYLE, Senior Program Officer
EVAN ELWELL, Research Associate
AMELIA GREEN, Senior Program Assistant
ADRIANNA HARGROVE, Finance Business Partner (through May 2021)
RYAN MURPHY, Program Officer
MARGUERITE SCHNEIDER, Administrative Coordinator (through February 2022)
DONAVAN THOMAS, Finance Business Partner (from June 2021)

---

[1] Member, National Academy of Engineering.
[2] Member, National Academy of Sciences.

Preface

To address what the 2018 National Defense Strategy describes as the "ever more lethal and disruptive battlefield, combined across domains, and conducted at increasing speed and reach," the U.S. Department of Defense (DoD) is pursuing an improved ability to more closely integrate and operate jointly against adversaries in a digital, distributed approach through Joint All-Domain Command and Control (JADC2).[1] To realize this concept will require the seamless integration of sensors, networks, platforms, commanders, operators, and weapon systems for rapid information collection, decision-making, and force projection.

The Department of the Air Force's (DAF's) contribution to JADC2 is the Advanced Battle Management System (ABMS), which seeks to modernize joint operations through sensor-to-shooter information collection, processing, routing, decision-making, and engagement to bring capabilities to bear faster against an agile adversary. Much attention has been given to ABMS, because it was presented as an evolving "system of systems"[2] and "a radically new acquisition model for the

---

[1] U.S. Department of Defense, 2018, *Summary of the 2018 National Defense Strategy of the United States of America: Sharpening the American Military's Competitive Edge*, Washington, DC.

[2] R. Uppal, 2021, "USAF Advanced Battle Management System (ABMS) Developing 'Internet of Military Things' to Enable Joint All-Domain Command and Control Comprising Family of Platforms Including Satellite Constellation," *International Defense, Security, & Technology (IDST) News*, https://idstch.com/space/usaf-advanced-battle-management-system-abms-developing-internet-of-military-things-to-enable-joint-all-domain-command-and-control-comprising-family-of-platforms-including-satellit/, March 30.

Air Force."[3] However, significant questions remain precisely because ABMS has not followed a traditional acquisition approach and the DAF projects that it will spend roughly $3.3 billion through fiscal year 2025.[4] Congress is therefore seeking greater clarity regarding ABMS's costs and technical development efforts.[5]

The Office of Management and Budget and the Department of the Air Force requested the National Academies of Sciences, Engineering, and Medicine to assess planned ABMS architecture, technology gaps, and governance. From October 2020 to May 2021, the Committee on Air Force Advanced Battle Management System conducted an extensive literature review from mostly open-source trade press and convened 12 unclassified sessions and one multi-day classified data gathering session to receive expert testimonies and collect information about available ABMS communications and systems integration architecture, technical approach, and governance structure plans and capabilities. Although the COVID-19 pandemic hampered the committee's ability to conduct site visits to operational and command and control (C2) centers, the committee was nonetheless able to collect valuable insights from the many experts who presented on ABMS and JADC2. The committee also held weekly virtual planning sessions from October 2020 to April 2021 and an in-person meeting in late May 2021 to deliberate and discuss key findings and recommendations. Writing commenced in June and was completed in September 2021.

The committee is grateful for the contributions of a wide range of noted experts and thought leaders to include representatives from the U.S. Departments of the Navy, the Army, and the Air Force regarding their respective communication systems and their approaches toward JADC2. Other expert organizations consulted during the course of the study included the Joint Staff, U.S. Northern Command, the Joint Artificial Intelligence Center, the National Security Agency, federally funded research and development centers, university-affiliated research centers, commercial industry, and numerous others. Many of the experts who participated

---

[3] A. McCullough, 2019, "ABMS Expected to Pick Up Speed with New Chief Architect in Place," *Air Force Magazine*, https://www.airforcemag.com/abms-expected-to-pick-up-speed-with-new-chief-architect-in-place/, March 10.

[4] R.S. Cohen, 2020, "Air Force Bets on ABMS Success in Fiscal 2021," *Air Force Magazine*, https://www.airforcemag.com/air-force-bets-on-abms-success-in-fiscal-2021/, February 11.

[5] See GAO (Government Accountability Office), 2019, "Defense Acquisitions: Action Is Needed to Provide Clarity and Mitigate Risks of the Air Force's Planned Advanced Battle Management System," https://www.gao.gov/assets/gao-20-389.pdf, April. See also Y. Tadjdeh, 2020, "Advanced Battle Management System Faces Headwinds," *National Defense Magazine*, https://www.nationaldefensemagazine.org/articles/2020/9/11/advanced-battle-management, September 11, and S.J. Freedberg, Jr., 2019, "House Armed Services Scrutinizes F-35 Costs, ABMS, Army Modernization," *Breaking Defense*, https://breakingdefense.com/2019/06/house-armed-services-scrutinizes-f-35-costs-abms-army-modernization/, June 3.

in the study's committee meetings have a distinguished record of public service, including in the military, and the committee thanks them for their service to our nation.

While ABMS remains an evolving ecosystem under development, this report summarizes the findings and recommendations of the National Academies consensus study on ABMS, providing a point-in-time perspective on what ABMS is and could be, and how it may be improved as it continues to evolve. This study was conducted by eight committee members and was greatly aided by our study director, Ellen Chou, and her excellent staff, including Evan Elwell and Ryan Murphy.

Philip S. Antón, *Chair*
Committee on Air Force Advanced Battle Management System

Acknowledgments

The committee would like to thank the following individuals for providing input to this study:

Andre' (Dre') B. Abadie, U.S. Army Futures Command
Christopher P. Azzano, Air Force Test Center, U.S. Air Force
Ally Bain, Office of Management and Budget
Marc Bernstein, Office of the Chief Architect, Department of the Air Force
Aaron Blow, MITRE Corporation
Eric Bryant, National Security Agency
Matthew Butkovic, Carnegie Mellon University Software Engineering Institute
Christopher Carey, MIT Lincoln Laboratory
Kelii "Koala" H. Chock, Air Force Life Cycle Management Center, U.S. Air Force
Dennis A. Crall, U.S. Marine Corps, Joint Staff J6, U.S. Department of Defense
Robert Cunningham, Carnegie Mellon University Software Engineering Institute
Mark Daniel, MIT Lincoln Laboratory
Preston Dunlap, Chief Architect, Department of the Air Force
Roy El-Rayes, Rapid Capabilities Office, Department of the Air Force
Drew Fanning, Chooch AI

James F. Geurts, U.S. Department of the Navy
Chad Haferbier, Rapid Capabilities Office, Department of the Air Force
Marin Halper, MITRE Corporation
Mark D. Happel, Johns Hopkins University Applied Physics Laboratory
George M. Hart III, Air Combat Command, U.S. Air Force
Walter C. Hattemer, Air Combat Command, U.S. Air Force
Christopher Hocking, Rapid Capabilities Office, Department of the Air Force
Lauren Knausenberger, Chief Information Officer, Department of the Air Force
Scott Lee, MITRE Corporation
Sherrill Lingel, RAND Corporation
Art Manion, Carnegie Mellon University Software Engineering Institute
John D. Matyjas, Air Combat Command, U.S. Air Force
Kelly McCool, Navy Digital Warfare Office, U.S. Department of the Navy
Dennis P. (Devo) McDevitt, Air Combat Command, U.S. Air Force
Keith C. McGuire, Air Combat Command, U.S. Air Force
Nicholas Miknev, Air Force Life Cycle Management Center, U.S. Air Force
Jeff J. Mrazik, DCS for Strategy, Integration and Requirements (AF/A5), U.S. Air Force
Nand Mulchandani, Joint Artificial Intelligence Center, U.S. Department of Defense
Aaron "Ocho" Nelson, Rapid Capabilities Office, Department of the Air Force
John "JP" A. Priestly III, Air Force Life Cycle Management Center, U.S. Air Force
Scott M. Roberts, Office of the Deputy Chief of Staff for Air Force Intelligence, Surveillance, Reconnaissance and Cyber Effects (AF/A2/A6), U.S. Air Force
Michele Schuman, MIT Lincoln Laboratory
Forrest Shull, Carnegie Mellon University Software Engineering Institute
Douglas W. Small, Naval Information Warfare Systems Command, U.S. Navy
John P. Stenbit, Viasat, Inc.
Katherine "Claire" Stowe, Air Force Life Cycle Management Center, U.S. Air Force
Matthew "Nomad" D. Strohmeyer, U.S. Air Force
Bryan Tipton, Rapid Capabilities Office, Department of the Air Force
Jeffery D. Valenzia, Office of the Deputy Chief of Staff for Strategy, Integration, and Requirements (AF/A5), U.S. Air Force
Kyle Volpe, Rapid Capabilities Office, Department of the Air Force

John Vona, Air Combat Command, U.S. Air Force
Randall "Waldo" Walden, Program Executive Officer, Rapid Capabilities Office, Department of the Air Force
Jennifer Watson, MIT Lincoln Laboratory
John S. Wellman, Joint Staff J6, U.S. Department of Defense
Martin Whelan, Aerospace Corporation
Stuart A. Whitehead, Joint Staff J6, U.S. Department of Defense

Acknowledgment of Reviewers

This Consensus Study Report was reviewed in draft form by individuals chosen for their diverse perspectives and technical expertise. The purpose of this independent review is to provide candid and critical comments that will assist the National Academies of Sciences, Engineering, and Medicine in making each published report as sound as possible and to ensure that it meets the institutional standards for quality, objectivity, evidence, and responsiveness to the study charge. The review comments and draft manuscript remain confidential to protect the integrity of the deliberative process.

We thank the following individuals for their review of this report:

Claude R. Canizares, NAS,[1] Massachusetts Institute of Technology,
Hao Huang, NAE,[2] University of Houston and University of Wisconsin,
Claire Leon, NAE, Loyola Marymount University,
Steven B. Lipner, NAE, SAFECode,
Leslie A. Momoda, HRL Laboratories, LLC,
Donald G. Sather, The Aerospace Corporation,
Patrick M. Shanahan, Former Deputy Secretary of Defense, U.S. Department of Defense,
Scott H. Swift, The Swift Group, LLC,

---

[1] Member, National Academy of Sciences.
[2] Member, National Academy of Engineering.

David M. Van Wie, NAE, Johns Hopkins University Applied Physics
   Laboratory, and
Stephen P. Welby, Institute of Electrical and Electronics Engineers.

Although the reviewers listed above provided many constructive comments and suggestions, they were not asked to endorse the conclusions or recommendations of this report nor did they see the final draft before its release. The review of this report was overseen by Alton D. Romig, NAE Executive Officer, Lockheed Martin Skunk Works (retired). He was responsible for making certain that an independent examination of this report was carried out in accordance with the standards of the National Academies and that all review comments were carefully considered. Responsibility for the final content rests entirely with the authoring committee and the National Academies.

Contents

SUMMARY 1

1 PERSPECTIVES 9
    Vision of Future Air and Space Operations, 12
    Joint All-Domain Command and Control (JADC2), 16
    Air Operations Center (AOC), 18
        Current AOC, 18
        An Alternative Future AOC, 22
    Advanced Battle Management System (ABMS), 23
        Evolution of ABMS, 23
        A Non-Traditional Acquisition Approach, 24
        From Demonstrations to Capabilities Releases, 28
        ABMS as a Contributor to JADC2, 30
        Other Contributors to JADC2 and Complicating Factors, 32

2 ARCHITECTURE AND DATA 36
    Architecture Overview, 36
        Architecture and Technology Status, 40
        Technology for Data-Centric Operations, 44
        Highly Capable Processing: AI and ML, 45
        Data and Data Standards, 48
        Containerization and Kubernetes, 50

  Software Considerations, 53
    Application Software and DevSecOps, 54
    Data Rights, 58
  Security, 59
    Network Reliability, Resiliency, and Fault Tolerance, 59
    Multi-Level Security, 62
    Cybersecurity and Zero Trust, 64
  Testing and Modeling, 68
    Test and Evaluation, 68
    Model-Based Systems Engineering, 70
    M&S and VV&A, 73
    Digital Twin, 74
  Common Mission Command Center, 76

3 GOVERNANCE   78
  Organization Integration, 82
  Human Factors, 86
    Human Systems Integration, 86
    Training, Culture, and Other Considerations, 88

4 CHALLENGES AND OPPORTUNITIES   93
  Interoperability, 93
  Intelligence, 95
  Major Recommendations, 96
    Technical, 96
    Non-Technical, 99
  Concluding Thoughts, 100

SELECTED BIBLIOGRAPHY   102

APPENDIXES

A Statement of Task   107
B Data-Gathering Meetings   109
C Acronyms and Abbreviations   116
D Committee Member Biographical Information   121
E Disclosure of Unavoidable Conflicts of Interest   128

Summary

The U.S. Department of Defense (DoD) is pursuing an improved ability to more closely integrate and operate jointly against agile adversaries through Joint All-Domain Command and Control (JADC2). This framework will seamlessly integrate sensors, networks, platforms, commanders, operators, and weapon systems for rapid information collection, decision-making, and projection of joint and multi-national forces. The Department of the Air Force's (DAF's) contribution to JADC2 is the Advanced Battle Management System (ABMS).

There are questions as to what ABMS is and concerns as to whether it is properly structured, because it lacks a well-defined set of discrete, allocated minimum performance objectives, a single set of fixed requirements, a timeline of proposed capability deliveries, and a systems allocation of budget and resources to execute against these objectives. While agility, flexibility, and adaptability are worthy goals, without a plan that offers sufficient details, specificity, and metrics to synchronize such a vast and complex system of systems approach, successful capability delivery at scale is challenged and unlikely.

To address these concerns, the committee was charged to examine the following:

1. Evaluate the planned ABMS data and communication architecture and compare the architecture anticipated performance characteristics needed to support real-time fire control and all-domain sensor-to-shooter data flow, command and control (C2) activities, artificial intelligence (AI)-based patterns-of-life training, battle damage assessment, and other related data-based activities;

2. Determine any technical gaps and shortfalls in ABMS technology and planned system integration architecture; and
3. Review ABMS governance and recommend how planned organizational and execution plans and processes may be improved to better enable a rapid realization of JADC2 operations for the DAF and DoD, as a whole.

In the conduct of the study, the former Assistant Secretary of the Air Force for Acquisition, Technology, and Logistics (SAF/AQ) transferred office of primary responsibility (OPR) for ABMS from the DAF's Chief Architect's Office (DAF CAO) to the DAF's Rapid Capabilities Office (RCO). The result of this change was twofold. First, the tasks that the committee was originally charged to examine did not fully align with the ABMS priorities and responsibilities of the RCO. For this reason, some of the information that the committee requested to complete the required analyses could not be provided and some of the information that the committee received was subsequently supplanted by newer information. Second, the committee received a largely transitory picture of ABMS, because both the system's technical design and governance were undergoing significant changes within the DAF.

As an evolving system in the early stages of definition, ABMS architecture and its supporting elements remain dynamic. The ABMS technical architecture presented to the committee from October 2020 to March 2021 largely reflected the status of ABMS emerging out of large-scale exercises known as "on-ramp" demonstrations managed by the DAF CAO. The early architecture and approach are undergoing assessments and revisions by the DAF RCO as it works to create a set of acquisition programs within the capability releases to be fielded. As such, the committee's analyses reflect the approach, benefits, challenges, and opportunities of that early architecture and constitute insights and recommendations for the CAO, RCO, DAF, and broader DoD elements to consider, as they pursue the updated ABMS architecture, the individual acquisition programs within it, and the larger JADC2 framework. The DAF RCO is already addressing some of these issues under the direction of the new Secretary of the Air Force in their evolving plans and designs for ABMS, but others (especially the non-technical elements) require further consideration.

REPORT ORGANIZATION AND MAJOR OBSERVATIONS

This report is organized by topic into four chapters: perspectives, architecture and data, governance, and challenges and opportunities. Chapter 1 describes why ABMS is needed and how it has evolved from a replacement program for a joint surveillance and radar system to an all-encompassing C2 family of systems. Chapter 2 examines current and planned architecture to include data standards, software, security, testing, and modeling. Chapter 3 outlines past and current governance for ABMS and highlights human systems integration, training, culture, and other considerations. Chapter 4 details interoperability and intelligence, and summarizes the committee's recommendations.

At a high level, the committee concludes that as a non-traditional acquisition program, ABMS is on track, but it remains a work in progress. Its technical design and architecture remain nascent and evolving, so it was difficult for the committee to conduct a comprehensive evaluation of its data and communication architecture, particularly as they relate to the JADC2 framework, which is also being developed and defined. Moreover, the committee found that performance characteristics were largely limited in scale and scope, because they were largely tied to on-ramp demonstrations and not actual operational activities, where real-world physical constraints may restrict actual performance.

The committee considers the assignment of the DAF RCO as the lead organization for ABMS to be a positive step toward shifting ABMS from demonstrations and experimentations to focused capability releases. The committee also supports the Secretary of the Air Force's call for the establishment of performance metrics to gauge improvements and measure operational outcomes.

As a family of systems, ABMS is difficult to quantify. The committee was not able to detail and assess the exact costs for ABMS, because it involves a portfolio of programs—some of which were not designated as elements of ABMS, but were still included as part of the broader ABMS ecosystem. Congress's decision to reduce the overall budget for ABMS by nearly one-half clearly limits what ABMS can accomplish in the near to mid-term. But this budget constraint may also compel DAF leaders to make imperative decisions and prioritizations regarding ABMS investments and capabilities, which the committee supports.

The committee found that the current ABMS, and the broader JADC2 governance structure, is insufficient and lacks proper authority to execute C2 across all domains. The absence of a DoD-level executive agent to address and resolve technical, operational, and command decisions for all contributors to the JADC2 framework leads to each Service and DoD agency developing its own C2 system, with unique requirements, standards, and technical specifications that challenge the achievement for interoperability.

The committee recognizes that ABMS has evolved in both its technical approach and its governance structure during the course of this analysis. It is thus important to note that some of the recommendations summarized below and detailed in the rest of this report are specific to the earlier ABMS approach, while others may remain relevant to a newer, more focused program.

RECOMMENDATIONS BY ORGANIZATION

The DAF's CAO and the RCO recommendations:

- Define the Advanced Battle Management System (ABMS) architecture at the Joint All-Domain Command and Control level to ensure interoperability with other ABMS-like systems being developed (Recommendation 1).
- Design the Advanced Battle Management System architecture to be modular and include open standards and interfaces that would enable configuration with other Service variants (Recommendation 3).
- Design the Advanced Battle Management System's architecture with specific technical requirements and solutions for ensuring that communications, data, and computation may continue to operate in degraded or denied access environments (Recommendation 4).
- To the maximum extent possible, design and execute a comprehensive artificial intelligence strategy that would encompass all elements, to include doctrine, chain of command, policy, authorization for weapon release in a joint environment, interfaces to Joint All-Domain Command and Control, and not just select capabilities of the Advanced Battle Management System (Recommendation 6).
- In coordination with the Department of the Air Force's Chief Software Officer, expand the use of containerization and Kubernetes for continuous Advanced Battle Management System development and for detecting and mitigating security vulnerabilities (Recommendation 8).
- Adopt development, security, and operations as the common development environment using containerization and continuous integration/continuous delivery across all of the Advanced Battle Management System (Recommendation 9).
- Design resilience into the Advanced Battle Management System architecture and specify dynamic criteria for needed performance (Recommendation 11).
- Work with the Department of the Air Force's Digital Engineering Enterprise Office to apply model-based systems engineering (MBSE) methods across Advanced Battle Management System engineering and sustainment activities and to enable MBSE to serve as a bridge between operator requirements and development teams (Recommendation 17).

The DAF's RCO recommendations:

- Adopt an array of data-exchange technologies that could support the entire spectrum of capabilities, from tactical to strategic (Recommendation 5).
- For modular open-system designs with robust interface specifications, acquire performance and interface requirements instead of all intellectual property rights (Recommendation 10).
- Apply zero trust (ZT) in stages as technologies mature and integrate ZT services to include the use of multi-factor authentication across all of the Advanced Battle Management System (Recommendation 13).
- In addition to adopting zero trust, leverage the best available mature cybersecurity practices and capabilities, including multi-factor authentication; identity, credential, and access management; encryption; penetration testing; managed detection services; behavior monitoring applications; among others (Recommendation 14).
- Employ the Air Force's Mission Defense Teams to red team the Advanced Battle Management System's cyber defenses against attacks from malicious actors. Based on these red team exercises, the Department of the Air Force Rapid Capabilities Office should address vulnerabilities by bolstering and enhancing cyber defenses accordingly (Recommendation 15).
- Work in partnership with the U.S. Cyber Command to address Internet of Things defense and other cyber vulnerabilities and exploits that are highlighted in the "United States Cyber Command Technical Challenge Problem Set" document (Recommendation 16).
- Building on existing activities in digital engineering and modeling and simulations, expand the use of digital twins in Advanced Battle Management System development, particularly as new capabilities and technologies are introduced (Recommendation 18).
- Consider scaling the Common Mission Control Center and designate it as phase zero for the Advanced Battle Management System (Recommendation 19).
- Incorporate human systems integration methodologies into the Advanced Battle Management System to ensure that all human users are fully and effectively integrated with current and future systems elements (Recommendation 22).

The Joint Staff, military services, and the U.S. DoD leaders' recommendations:

- Establish interoperability requirements and performance metrics for all participants in Joint All-Domain Command and Control to allow for eventual integration of all capabilities (Recommendation 2).
- Reach immediate agreement on a common data fabric and security levels of the data with data standards and tools defined at the Joint level. Without a common set of agreed upon open standards with known interface exchange requirements that do not limit innovation, the military Services risk developing incompatible and stove-piped solutions (Recommendation 7).
- Establish and implement a robust enterprise-wide offensive and defensive cybersecurity strategy for Joint All-Domain Command and Control (JADC2) and the Advanced Battle Management System. Security is a fundamental requirement that must be designed and fully integrated into the all JADC2-supporting systems' architecture from the start (Recommendation 12).
- Establish an authoritative Joint-level body to address and resolve technical, operational, and command decisions for all contributors to the Joint All-Domain Command and Control framework (Recommendation 20).
- Tackle the cultural, social, and emotional barriers to true Joint Warfighting Concept (JWC) horizontal integration if the Advanced Battle Management System and the larger Joint All-Domain Command and Control constructs are to enable the truly joint and multi-national integrated operations envisioned by the JWC (Recommendation 21).
- Ensure that the ethical use of artificial intelligence is examined and addressed in the Advanced Battle Management System's (and in other systems supporting the broader Joint All-Domain Command and Control framework's) design, operation, staffing, and training, as dictated by policy and the law of war (Recommendation 24).

The DAF's SAF/AQ and the Deputy Chief of Staff for Strategy, Integration, and Requirements recommendation:

- Consider and weave personnel, cultural, training, and other non-materiel doctrine, organization, training, materiel, leadership, education, personnel, facilities, and policy issues into designs and implementation plans for the broader Advanced Battle Management System ecosystem (Recommendation 23).

The Air Education Training Command recommendation:

- Establish a curriculum that would train or recruit highly qualified experts in artificial intelligence/machine learning, model-based systems engineering, cybersecurity, intelligence assessment, and test and evaluation for information technology, software, and hardware who can work with experts in military operations and culture (Recommendation 25).

Each of these recommendations is explored in detail in the full report.

1

Perspectives

Our Air Force must accelerate change to control and exploit the air domain to the standard the Nation expects and requires from us. If we don't change—if we fail to adapt—we risk losing the certainty with which we have defended our national interests for decades. We risk losing a high-end fight. We risk losing quality Airmen, our credibility, and our ability to secure our future. We must move with a purpose—we must Accelerate Change or Lose.
——General Charles Q. Brown, Jr., Chief of Staff, U.S. Air Force[1]

To accelerate change to control and exploit the air domain, the Advanced Battlefield Management System (ABMS)[2] is the Department of the Air Force's

---

[1] C.Q. Brown, Jr., 2020, *Accelerate Change or Lose*, https://www.af.mil/Portals/1/documents/2020SAF/ACOL_booklet_FINAL_13_Nov_1006_WEB.pdf, August.

[2] P. Dunlap, 2020, "ABMS Overview," Presentation to the Air Force ABMS Committee, October 30. See also J. Eddins, 2021, "Valenzia: ABMS Will Deliver the 'Decision Advantage,'" *Airman Magazine*, https://www.airmanmagazine.af.mil/Features/Display/Article/2634972/valenzia-abms-will-deliver-the-decision-advantage/, May 26. D. Mayer. 2021, "ABMS Aims to Revolutionize Data Flow, Speed Decisions," *Air Force News*, April 1, https://www.af.mil/News/Article-Display/Article/2559022/abms-aims-to-revolutionize-data-flow-speed-decisions/. CRS (Congressional Research Service), 2021, *Advanced Battle Management System (ABMS)*, , https://sgp.fas.org/crs/weapons/IF11866.pdf, June 29.

(DAF's) contribution to the Joint All-Domain Command and Control (JADC2)[3] concept of seamless joint and multi-national information sharing and operational command and control (C2). These efforts seek to allow current and future sensors, commands, operators, and weapon systems to share appropriate and accurate information at the speeds required to overcome anticipated adversary's decision-making and actions.[4] ABMS seeks to share critical operational data across the U.S. Department of Defense (DoD) enterprise in both contested high-end and low intensity warfighting environments. Information such as intelligence, surveillance, and reconnaissance (ISR) on the threat, weapons systems availability, and military operational status and actions underpin the ABMS architecture to allow joint and allied commanders to achieve an accurate, real-time understanding of the environment and take action faster than any potential adversary's observe-orient-decide-act (OODA) loop.[5]

Owing to the evolving nature of ABMS, there is some confusion as to what it actually is today and promises to be tomorrow.[6] Originating from other concepts known as Multi-Domain or All-Domain C2 and the Air Force's cancelled Joint

---

[3] See J. Garamone, 2020, "Joint All-Domain Command, Control Framework Belongs to Warfighters," *DoD News*, November 30, https://www.defense.gov/Explore/News/Article/Article/2427998/joint-all-domain-command-control-framework-belongs-to-warfighters/. CRS, 2021, *Joint All-Domain Command and Control: Background and Issues for Congress*, https://crsreports.congress.gov/product/pdf/R/R46725/2, March 18. CRS, 2021, *Joint All-Domain Command and Control (JADC2)*, https://sgp.fas.org/crs/natsec/IF11493.pdf, July 1.

[4] Office of the Under Secretary of Defense (Comptroller/Chief Financial Officer), *Defense Budget Overview, United States Department of Defense Fiscal Year 2022 Budget Request*, 2021, https://comptroller.defense.gov/Portals/45/Documents/defbudget/FY2022/FY2022_Budget_Request_Overview_Book.pdf, May.

[5] See "DoD's Data-Driven Future: Shared Knowledge, Near Real-Time Answers," 2021, *Air Force Magazine*, https://www.airforcemag.com/dods-data-driven-future-shared-knowledge-near-real-time-answers/, June 1. G.S. Fein, 2003, "New Meaning for 'OODA Loop,'" *National Defense Magazine*, https://www.nationaldefensemagazine.org/articles/2003/10/1/2003october-new-meaning-for-ooda-loop.

[6] See T. Hitchens, 2020, "Roper Mulls Name Change for Changing ABMS (Not Skynet!)," *Breaking Defense*, https://breakingdefense.com/2020/09/roper-mulls-name-change-for-changing-abms-not-skynet/, September 4. S. Sirota, 2021, "Roper Caves to Demands, Establishing ABMS as a Traditional Acquisition Program," *Inside Defense*, https://insidedefense.com/daily-news/roper-caves-demands-establishing-abms-traditional-acquisition-program, January 14. M. Mayfield, 2020, "Air Force's Advanced Battle Management System Takes New Step," *NDIA Magazine*, https://www.nationaldefensemagazine.org/articles/2020/11/24/advanced-battle-management-system-takes-new-step, November 24. T. Hitchens, 2020, "Roper Targets Commercial AI, Data Analytics for Next ABMS Deals," *Breaking Defense*, https://breakingdefense.com/2020/05/roper-targets-commercial-ai-data-analytics-for-next-abms-deals/, May 14.

Surveillance and Target Attack Radar System (JSTARS) recapitalization program,[7] ABMS is both a concept and an initiative to field wide-ranging capabilities. ABMS has become synonymous with JADC2, but that misleadingly lacks consideration of other JADC2 efforts in the Army and Navy.[8] Moreover, ABMS (and JADC2) will require adjustments not only in materiel but across the entire spectrum known as *DOTMLPF-P*—that is, DoD will have to address doctrine, organization, training, materiel, leadership, education, personnel, facilities, and policy (DOTMLPF-P)[9] aspects to successfully implement ABMS. Thus, while ABMS will visibly field capabilities and systems, it is envisioned to impact the *totality* of military operations across the board. This is why various organizations have a hard time understanding what ABMS is and what it may become. From concept to execution, ABMS seeks to focus not only on C2, networks, and specific weapons systems and platforms, but also on an enterprise-wide solution to connect and integrate all force capabilities. Figure 1.1 depicts the DAF's vision of ABMS.

To provide perspectives on the potential for ABMS and its evolving efforts, this chapter first offers a foresight of what future air and space operations might entail under ABMS and JADC2, followed by an overview of recent and current activities on what ABMS is, its motivation, expectations, timelines, and relationships with other Service elements of JADC2.

---

[7] See A. McCullough, 2019, "Life After JSTARS," *Air Force Magazine*, https://www.airforcemag.com/article/life-after-jstars/, March 21. S.J. Freedberg, Jr., 2019, "Air Force ABMS: One Architecture to Rule Them All?" *Breaking Defense*, https://breakingdefense.com/2019/11/air-force-abms-one-architecture-to-rule-them-all/, November 8.

[8] A. Abadie, 2021, "Project Convergence Overview," Presentation to the Air Force ABMS Committee, January 8. D.W. Small, 2021, "Project Overmatch," Presentation to the Air Force ABMS Committee, March 3. See also K. Underwood and R.K. Ackerman, 2021, "Services Choose Independent Paths for JADC2," *SIGNAL*, https://www.afcea.org/content/services-choose-independent-paths-jadc2, April 1. T. Hitchens, 2021, "Combatant Commands Worry About Service JADC2 Stovepipes," *Breaking Defense*, https://breakingdefense.com/2021/08/combatant-commands-worry-about-service-jadc2-stovepipes/?utm_campaign=Breaking%20News&utm_medium=email&_hsmi=154724449&_hsenc=p2ANqtz-_wkEKLecBcHxjyqNV8fxrrXZamBG1AweYP8ls7P6GEAnGBkbZL9XpBOgj5Ks7w7gCxJtCsOKaC9947kj0GT15Y2tliBA&utm_content=154724449&utm_source=hs_email, August 31.

[9] For a definition on DOTMLPF-P, see Chairman of the Joint Chiefs of Staff Instruction (CJCSI), 2016, CJCSI 3010.02E, *Guidance for Developing and Implementing Joint Concept*, pp. A-3–A-5, https://www.jcs.mil/Portals/36/Documents/Library/Instructions/CJCSI%203010.02E.pdf, August 17. For more on the spectrum of ABMS requirements, see D. Allvin, 2021, "Why We Need the Advanced Battle Management System," *DefenseOne*, https://www.defenseone.com/ideas/2021/05/why-we-need-advanced-battle-management-system/173861/, May 6.

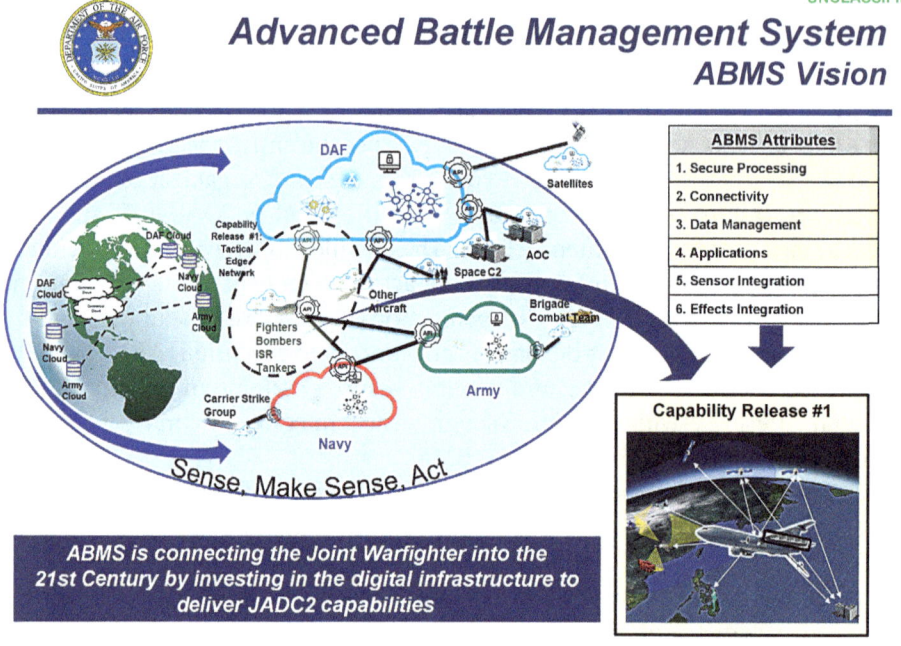

FIGURE 1.1 ABMS Vision. SOURCE: Randall G. Walden, Air Force Rapid Capabilities Office, Department of the Air Force. Presentation to the committee on January 22, 2021. Approved for public release.

VISION OF FUTURE AIR AND SPACE OPERATIONS

The *2018 National Defense Strategy* (NDS) describes "an increasingly complex global security environment, characterized by overt challenges to the free and open international order and the reemergence of long-term, strategic competition between nations.… We face an ever more lethal and disruptive battlefield, combined across domains, and conducted at increasing speed and reach—from close combat, throughout overseas theaters, and reaching to our homeland. Some competitors and adversaries seek to optimize their targeting of our battle networks and operational concepts, while also using other areas of competition short of open warfare to achieve their ends (e.g., information warfare, ambiguous or denied proxy operations, and subversion)."[10] Notably,

---

[10] DoD (U.S. Department of Defense), 2018, *Summary of the 2018 National Defense Strategy of the United States of America: Sharpening the American Military's Competitive Edge*, https://dod.defense.gov/Portals/1/Documents/pubs/2018-National-Defense-Strategy-Summary.pdf.

the security environment is also affected by rapid technological advancements and the changing character of war. The drive to develop new technologies is relentless, expanding to more actors with lower barriers of entry, and moving at accelerating speed. New technologies include advanced computing, "big data" analytics, artificial intelligence, autonomy, robotics, directed energy, hypersonics, and biotechnology—the very technologies that ensure we will be able to fight and win the wars of the future. New commercial technology will change society and, ultimately, the character of war.[11]

As a result, "It's really about who can sense and make sense of their environment and take action faster than their opponent. Victories come to the side that can decide quickly and accelerate that kill chain. We call and seek that *decision advantage*."[12]

The subsequent issuance of DoD's *Digital Modernization Strategy*[13] in 2019 highlighted the DoD's recognition that to address these pressing challenges, current technologies need to rapidly advance into a digital future. As modern battlefields shift toward farther, distributed, and progressively complex interconnected warfighting domains, ensuring communication, coordination, and execution becomes increasingly more important. Ensuring that forces in space, cyberspace, air, land, surface, and subsurface can effectively and promptly communicate to support both kinetic and non-kinetic operations is critical. This explosion of the digital era over the past two decades—with particular emphasis on artificial intelligence (AI) and pervasive communication and processing technologies—has already changed and is expected to continue changing the battlespace dramatically.

For example, as stated in the final report from the National Security Commission on Artificial Intelligence (NSCAI):

> AI is expanding the window of vulnerability the United States has already entered. For the first time since World War II, America's technological predominance—the backbone of its economic and military power—is under threat. China possesses the might, talent, and ambition to surpass the United States as the world's leader in AI in the next decade if current trends do not change. Simultaneously, AI is deepening the threat posed by cyber

---

[11] DoD (U.S. Department of Defense), 2018, *Summary of the 2018 National Defense Strategy of the United States of America: Sharpening the American Military's Competitive Edge*, https://dod.defense.gov/Portals/1/Documents/pubs/2018-National-Defense-Strategy-Summary.pdf, p. 3.

[12] J. Valenzia, 2020, "Joint Warfighting Concept: Joint All Domain Command and Control (JADC2) and the Advanced Battle Management System (ABMS)," Presentation to the Air Force ABMS Committee, December 18.

[13] DoD, 2019, *DoD Digital Modernization Strategy*, https://media.defense.gov/2019/Jul/12/2002156622/-1/-1/1/DOD-DIGITAL-MODERNIZATION-STRATEGY-2019.PDF, July 12.

attacks and disinformation campaigns that Russia, China, and others are using to infiltrate our society, steal our data, and interfere in our democracy.[14]

Within Russia, "the development and use of AI is [viewed as] essential to the future success of Russia's armed forces and key to its military power.... Russian military strategists have placed a premium on establishing what they refer to as 'information dominance on the battlefield,' and AI-enhanced technologies promise to take advantage of the data available on the modern battlefield to protect Russia's own forces and deny that advantage to the adversary."[15]

Similarly, "Chinese military initiatives in AI are motivated by an acute awareness of global trends in military technology and operations ... and recognition of potential opportunities inherent in this military and technological transformation."[16] For this reason, Chinese military and China's defense industry have been pursuing significant investments in robotics, autonomy, and other applications of AI.[17] DoD's 2020 *Annual Report to Congress on China's Military and Security Developments* quotes from China's own *Next Generation AI Plan* that the country seeks to gain parity with global leaders in AI by 2020, achieving major breakthroughs in AI by 2025, and establishing China as the global leader in AI by 2030. "The PRC [People's Republic of China] is pursuing a whole-of-society effort to become a global leader in AI, which includes designating select private AI companies in China as 'AI champions' to emphasize research and development (R&D) in specific dual-use technologies."[18]

Compounding to the use of AI, adversaries also seek to undermine U.S. dominance in all domains through the use of hybrid warfare. While not a new concept, hybrid warfare, in which an adversary "simultaneously and adaptively employs a tailored mix of conventional, irregular, terrorism, and criminal means or activities in the operational battle space," including the use of non-kinetic tools to destabilize nations, has expanded the implements needed to manage the spectrum of warfare

---

[14] NSCAI (National Security Commission on Artificial Intelligence), 2021, *National Security Commission on Artificial Intelligence Final Report*, p. 7, https://www.nscai.gov/wp-content/uploads/2021/03/Full-Report-Digital-1.pdf, March.

[15] J. Edmonds, S. Bendett, A. Fink, et al., 2021, "Artificial Intelligence and Autonomy in Russia," Center for Naval Analyses, https://www.cna.org/CNA_files/centers/CNA/sppp/rsp/russia-ai/Russia-Artificial-Intelligence-Autonomy-Putin-Military.pdf, May.

[16] E.B. Kania, 2020, " 'AI Weapons' in China's Military Innovation," Brookings Institution, https://www.brookings.edu/wp-content/uploads/2020/04/FP_20200427_ai_weapons_kania_v2.pdf, April.

[17] E.B. Kania, 2020, " 'AI Weapons' in China's Military Innovation," Brookings Institution, https://www.brookings.edu/wp-content/uploads/2020/04/FP_20200427_ai_weapons_kania_v2.pdf, April, p. 1.

[18] DoD, 2020, *Annual Report to Congress: Military and Security Developments Involving the People's Republic of China*, https://media.defense.gov/2020/Sep/01/2002488689/-1/-1/1/2020-DOD-CHINA-MILITARY-POWER-REPORT-FINAL.PDF.

beyond conventional forces.[19] "Our strategic competitors have studied how we fight and they have taken asymmetric steps to exploit our vulnerabilities and to defeat us. We have to respond with a sense of urgency, but we also have to take the time necessary to make smart choices about our future and our investments."[20] As noted by the Ministry of Foreign Affairs of Ukraine:

> Military aggression is just one element of the Russian hybrid warfare against Ukraine. Other elements encompass: 1) propaganda based on lies and falsifications; 2) trade and economic pressure; 3) energy blockade; 4) terror and intimidation of Ukrainian citizens; 5) cyber attacks; 6) a strong denial of the very fact of war against Ukraine despite large scope of irrefutable evidence; 7) use of pro-Russian forces and satellite states in its own interests; 8) blaming the other side for its own crimes.[21]

Additionally, command and control (C2) timelines for protection against, and delivery of, "fast" weapons (e.g., hypersonic missiles) have also altered actionable timelines, cutting down not only on available timelines to conduct decisions, but also on the ability to calculate optimal options for defense. Technology and operational changes (e.g., AI, unmanned platforms, and the new "battlefield" of contested space) have driven the military Services (both individually and jointly) to reconsider what the operational C2 concept—and associated technological means to achieve it—should be for the future. Coupled with the new reality that the United States will be a smaller force, the nation will no longer have the luxury afforded by having both the largest and most technically advanced force in the field. All of these collective challenges undermine U.S. information dominance, which ABMS seeks to overcome. For the joint community, the new approach is JADC2 serving the Joint Warfighting Concept (JWC).[22]

---

[19] R.W. Glenn, 2009, "Thoughts on 'Hybrid' Conflict," *Small Wars Journal*, p. 2, https://smallwarsjournal.com/blog/journal/docs-temp/188-glenn.pdf, March 2. See also B.P. Fleming, 2011, *The Hybrid Threat Concept: Contemporary War, Military Planning and the Advent of Unrestricted Operational Art*. School of Advanced Military Studies, U.S. Army Command and General Staff College, https://cgsc.contentdm.oclc.org/digital/collection/p4013coll3/id/2752/.

[20] J. Tirpak, 2021, "Kendall: Modernize Now to Counter China," *Air Force Magazine*, https://www.airforcemag.com/kendall-modernize-now-to-counter-china/, September 20.

[21] Ministry of Foreign Affairs of Ukraine, 2019, "Ten Facts You Should Know About Russian Military Aggression Against Ukraine," https://mfa.gov.ua/en/10-facts-you-should-know-about-russian-military-aggression-against-ukraine, December 19.

[22] See T. Greenwood and P. Savage, 2019, "In Search of a 21st Century Joint Warfighting Concept," *War on the Rocks*, https://warontherocks.com/2019/09/in-search-of-a-21st-century-joint-warfighting-concept/, September 12. DoD, 2020, "Mission Engineering: Ensuring Key Technologies Drive the Joint Warfighting Concept," https://www.defense.gov/Explore/News/Article/Article/2391597/mission-engineering-ensuring-key-technologies-drive-the-joint-warfighting-conce/, October 22.

JOINT ALL-DOMAIN COMMAND AND CONTROL (JADC2)

JADC2 is the "warfighting capability to sense, make sense, and act at all levels and phases of war, across all domains, and with partners, to deliver information advantage at the speed of relevance."[23] It is DoD's solution to connect sensors from all of the military services—Air Force, Army, Marine Corps, Navy, and Space Force—into a single network.[24] JADC2 is "about creating a resilient, adaptable line of communication (e.g., mesh network) across the entire Joint Force, at every echelon, from the strategic level to the tactical edge. That protected and hardened network will power the ubiquitous flow of relevant information across all domains around the globe, enabling our warfighting commanders and senior leaders to make decisions and direct actions better and faster than our adversaries—to deter their actions and intents, if at all possible, and defeat them outright when necessary."[25] Figure 1.2 depicts the Foundations of JADC2.

Traditionally, each military Service has developed its own C2 network that is unique and generally incompatible across weapons systems, platforms, and operating domains. As a result, decision time cycles and the transmission of time-sensitive data to inform decisions were slow, at times duplicative, and organizationally stove-piped. JADC2 is DoD's enterprise-solution to this technical and operational challenge and "envisions providing a cloud-like environment for the Joint Force to share intelligence, surveillance, and reconnaissance data, transmitting across many communications networks, to enable faster decision-making."[26] Furthermore, it seeks to reimagine headquarter elements such as Air Operations Centers (AOCs) that are commonly removed from the operating battlespace, and empower forward-deployed combat commanders with the same situational awareness and decision-making as operations HQ. "This concept enables force management that is responsive to, even out in front of, enemy or adversary generated effects, decision-making, and maneuvering."[27]

Throughout history, victory often goes to the entity with an ability to make decisions faster than its adversary and thus act more appropriately to capitalize on

---

[23] DoD, 2021, "Fact Sheet on JADC2." https://insidedefense.com/sites/insidedefense.com/files/documents/2021/jun/06042021_jadc2.pdf, June 4.

[24] CRS, 2021, "Joint All-Domain Command and Control (JADC2)," https://crsreports.congress.gov/product/pdf/IF/IF11493, July 1.

[25] DoD, 2021, "Fact Sheet on JADC2."

[26] Ibid.

[27] B.M. Pirolo, 2020, "Information Warfare and Joint All-Domain Operations," *Air & Space Power Journal*, 34(4):104.

FIGURE 1.2 Foundations of Joint All-Domain Command and Control (JADC2). SOURCE: *Air Force Magazine*, October 28, 2020.

previous decisions.[28] Moreover, the force with better situational awareness often dominates.[29] The quantity and fighting capability of forces are other major factors, but superior awareness and responsiveness are proven force multipliers. That does not mean such awareness will be perfect at all times, but a force with a superior ability to continue seeing, deciding, and fighting in degraded environments has a

---

[28] See J. Dransfield, 2020, "How Relevant Is the Speed of Relevance?: Unity of Effort Towards Decision Superiority Is Critical to Future U.S. Military Dominance," *The Bridge*, https://thestrategybridge.org/the-bridge/2020/1/13/how-relevant-is-the-speed-of-relevance-unity-of-effort-towards-decision-superiority-is-critical-to-future-us-military-dominance, January 13. "Giving Airmen the Edge: The Promise of JADC2," 2020, *Air Force Magazine*, https://www.airforcemag.com/giving-airmen-the-edge-the-promise-of-jadc2/, October 28.

[29] See National Academies of Sciences, Engineering, and Medicine, 2021, *Adapting to Shorter Time Cycles in the United States Air Force: Proceedings of a Workshop Series*, The National Academies Press, Washington, DC. M.C. Libicki and S.E. Johnson, editors, 1995, *Dominant Battlespace Knowledge*. National Defense University, http://www.dodccrp.org/files/Libicki_Dominant.pdf, October. M.W. Jones, 2020, "Strategic Decision Making—A Case Study," *Military Strategy Magazine*, 7(2):20–24, https://www.militarystrategymagazine.com/article/strategic-decision-making-a-case-study.

greater advantage toward achieving success. JADC2 is presented as the joint solution that would "synchronize the prosecution of thousands of potential targets across a federated resource set of the combat arms inherent to the task force and across domains"[30] and thus "provide the U.S. military with decision advantage in a future conflict with China or Russia by enabling U.S. forces to understand better, decide smarter, and act faster than adversaries."[31]

For the DAF, ABMS is the DAF's contribution to JADC2. "As a new approach toward information sharing and decision management, ABMS enables compressed decision-making and converging effects without domain or geographic boundaries … this speed matters to the decision maker and the warfighter. And, with the proliferation of technology, future warfighters will have the ability to observe, orient, decide, and act within minutes—as opposed to hours and days."[32]

The C2 functions that the DAF must perform—currently centralized in the AOC—must equally adapt to this accelerated decision-making environment with technologies that meet these demands. Because ABMS is presented as the new AOC materiel solution to highlight this change, the following provides a brief overview of the current AOC along with a possible view of what the AOC may become, given the promise of ABMS and its associated technologies.

AIR OPERATIONS CENTER (AOC)

Current AOC

The current AOC—with its current basic architecture originally designed at the beginning of the 21st century—is "both an Air Force unit and a Weapon System.… [It] is the [Joint or Combined Forces Air Component Commander's (JFACC)] C2 center that provides the capability to plan, direct, and assess activities of assigned and attached forces … and provides operational-level C2 of air, space, cyberspace and [information operations] IO to meet JFACC operational objectives and

---

[30] B.M. Pirolo, 2020, "Information Warfare and Joint All-Domain Operations," p. 104.

[31] Govini, 2021, *Department of Defense Investments in Joint All-Domain Command & Control Taxonomy*, https://govini.com/wp-content/uploads/2021/09/DoD-Investments-in-JADC2-Taxonomy.pdf.

[32] J.P. Roth and C.Q. Brown, Jr., 2021, "Department of the Air Force Posture Statement Fiscal Year 2022," Presentation to the Committees and Subcommittees of the U.S. Senate and the House of Representatives, 1st Session, 117th Congress, https://www.armed-services.senate.gov/imo/media/doc/FY22%20DAF%20Posture%20Statement%20-%20Final%20(v23.1)1.pdf.

guidance."[33] The AOC serves "as the focal point for designing, planning, executing, and assessing air component operations."[34]

The primary functions of the AOC are:

- Develop air component operations strategy and planning documents that integrate air, space, and cyberspace operations to meet air component commander objectives and guidance the Joint Force Commander (JFC) designates.
- Task, execute, and assess day-to-day air component operations; provide rapid reaction, positive airspace control, and coordinate and de-conflict weapons employment as well as integrate the total air component effort.
- Receive, assemble, analyze, filter, and disseminate all-source intelligence and weather information to support air component operations planning, execution, and assessment.
- Integrate space capabilities and coordinate space activities for the air component commander when designated as space coordinating authority.
- Issue airspace control procedures and coordinate airspace control activities for the airspace control authority (ACA) when designated.
- Provide overall direction of air defense, including theater missile defense (TMD), for the Area Air Defense Commander (AADC) when designated.
- Plan, task, and execute the theater air component intelligence, surveillance, and reconnaissance (ISR) mission.
- Conduct component-level assessment to determine mission and overall air component operations effectiveness as required by the JFC to support the theater assessment effort.
- Plan and task air mobility operations according to the theater priorities.[35]

As a weapon system, the current Air Operations Center—Weapon System (AOC-WS), known as the AN/USQ-163 Falconer, is "a system of systems that incorporates numerous third-party software applications and commercial off-the-shelf products. Each third-party system integrated into the AOC-WS provides its

---

[33] DAF (Department of the Air Force), 2020, *Department of the Air Force Manual 13-1AOC, Volume 3, Nuclear Space, Missile Command and Control Operational Procedures—Air Operations Center (AOC) Operations Center (OC)*, https://static.e-publishing.af.mil/production/1/af_a3/publication/dafman13-1aocv3/dafman13-1aocv3.pdf, December 18.

[34] Curtin E. LeMay Center, 2020, *Air Force Doctrine Publication (AFPD) 3-30: Appendix B: The Air Operations Center*, https://www.doctrine.af.mil/Portals/61/documents/AFDP_3-30/3-30-D70-C2-Appendix-AOC.pdf, January 7.

[35] Ibid.

own programmatic documentation. AOC-WS capabilities include command and control (C2) of joint theater air and missile defense; preplanned, dynamic, and time-sensitive multi-domain target engagement operations; and intelligence, surveillance, and reconnaissance operations management."[36] Additionally, the AOC-WS consists of:

- Commercial off-the-shelf software and hardware for voice, digital, and data communications infrastructure.
- Government software applications developed specifically for the AOC-WS to enable planning, monitoring, and directing the execution of air, space, and cyber operations, to include:
  - Additional third-party systems that accept, process, correlate, and fuse C2 data from multiple sources and share them through multiple communications systems.
- When required, the AOC-WS operates on several different networks, including the secret Internet protocol router network (SIPRNET), Joint Worldwide Intelligence Communications System, and coalition networks. The networks connect the core operating system and primary applications to joint and coalition partners.[37]

Currently, the Air Force's Kessel Run Experimentation Lab (KREL) is responsible for developing and deploying the AOC-WS Block 20 software to the field.[38] The goal is to modernize the AOC to enable a distributed AOC weapon system and to deprecate the existing 10.1 Falconer Weapon System.[39] As an integrated partner with operators, Kessel Run developers understand user needs and are able to develop and test software to meet those operational needs. If successful, this model of embedding software developers with end users should be adopted more broadly.

While progress is being made, notable challenges remain with the current AOC design and construct. Specifically, the underlying architecture of the DAF C2 functions in the AOC is not adequately designed to meet current operational and technological threats or support an accelerated pace of planning. As noted in a recent RAND study on JADC2, "the cancellation of the AOC 10.2 moderniza-

---

[36] R.F. Behler, 2020, *Director, DoD Operational Test and Evaluation Fiscal Year 2020 Annual Report*, https://www.dote.osd.mil/Publications/Annual-Reports/2020-Annual-Report, January.

[37] R.F. Behler, 2020, *Director, DoD Operational Test and Evaluation Fiscal Year 2020 Annual Report*, https://www.dote.osd.mil/Publications/Annual-Reports/2020-Annual-Report, January, p. 179.

[38] The AOC-WS Block 20 is a middle tier of acquisition (MTA) program intended to replace AOC-WS 10.1 with modernized, integrated, automated, and redundant capabilities to meet valid requirements defined for the previously canceled AOC-WS 10.2 program. Reference: see note 31.

[39] B. Katz and P. Ising, 2021, "Kessel Run Deploys KRADOS to Air Operations Center," *Kessel Run News*, https://kesselrun.af.mil/news/Kessel-Run-Deploys-KRADOS.html, January 12.

tion effort has delayed the delivery of critical hardware and software upgrades to the AOC … [and] growing emphasis on improved cyber and space integration has placed new functional and technical demands on the AOC and increased interest in multidomain operations."[40] As a result, "the Air Force AOC 72-hour air-tasking cycle is incongruent with the current digital world."[41]

The current AOC-WS program has also been historically challenged in the prioritization for funding. It has been impacted by the absence of a Joint Architecture and AI policy guideline toward which to build. The testimonies the committee received from operators and acquisition personnel highlighted the concerns of the current AOC not being aligned to the new JADC2, JWC, and threat needs. It was clear that the AOC system of systems architecture (as currently constructed) would not support a transformation over time, because the inherently outdated technology and architecture utilized by the current system is unable to be restructured.[42]

It is moreover evident—even without a JADC2 or JWC—that the Air Force requires an innovative and revamped AOC to interoperate with the new U.S. Space Command (USSPACECOM) and U.S. Space Force (USSF) operating systems and to meet broad operational challenges from adversaries seeking to counter U.S. military advantages through anti-access/area denial (A2/AD) from electronic warfare, cyber weapons, long-range missiles, advanced air defenses, and potentially even GPS-denial.[43] All these increase the need for faster decisions that leverage and integrate all U.S. military capabilities.

---

[40] S. Lingel, J. Hagen, E. Hastings, et al., 2020, "Joint All-Domain Command and Control for Modern Warfare," RAND Corporation, https://www.rand.org/content/dam/rand/pubs/research_reports/RR4400/RR4408z1/RAND_RR4408z1.pdf. See also S. Lingel, 2021, "ABMS and JADC2," Presentation to the Air Force ABMS Committee, April 21.

[41] S. Lingel, J. Hagen, E. Hastings, et al., 2020, "Joint All-Domain Command and Control for Modern Warfare," RAND Corporation, https://www.rand.org/content/dam/rand/pubs/research_reports/RR4400/RR4408z1/RAND_RR4408z1.pdf, p. viii. See also DoD, 2019, *Joint Publication 3-30, Joint Air Operations*, https://www.jcs.mil/Portals/36/Documents/Doctrine/pubs/jp3_30.pdf, July 25.

[42] Anthologized from various presentations given by DAF representatives to the Air Force ABMS Committee from March 30 to March 31, 2021. See also CRS, 2021, "Joint All-Domain Command and Control (JADC2)."

[43] See C. Dougherty, 2020, "Moving Beyond A2/AD," Center for New American Security, https://www.cnas.org/publications/commentary/moving-beyond-a2-ad, December 3. N. Impson, 2020, "The Next Warm War: How History's Anti-Access/Area Denial Campaigns Inform the Future of War," *Small Wars Journal*, https://smallwarsjournal.com/jrnl/art/next-warm-war-how-historys-anti-accessarea-denial-campaigns-inform-future-war, January 14. A. Krepinevich, B. Watts, and R. Work, 2003, "Meeting the Anti-Access and Area-Denial Challenge," Center for Strategic and Budgetary Assessments, https://csbaonline.org/uploads/documents/2003.05.20-Anti-Access-Area-Denial-A2-AD.pdf.

An Alternative Future AOC

Rapid advancements in technology have dramatically altered operational requirements and shortened response times in confronting adversarial threats. The *2018 National Defense Strategy* states, "This increasingly complex security environment is defined by rapid technological change, challenges from adversaries in every operating domain, and the impact on current readiness from the longest continuous stretch of armed conflict in our Nation's history. In this environment, there can be no complacency—we must make difficult choices and prioritize what is most important to field a lethal, resilient, and rapidly adapting Joint Force. America's military has no preordained right to victory on the battlefield."[44] The Chief of Staff of the U.S. Air Force, echoed this view when he stated, "While the Nation was focused on countering violent extremist organizations, our competitors focused on defeating us. They have studied, resourced, and introduced systems specifically designed to defeat the U.S. Air Force capabilities that have underpinned the American way of war for a generation.... In an environment that includes, but is not limited to, declining resources, aggressive global competitors, and rapid technology development and diffusion, the U.S. Air Force must accelerate change to control and exploit the air domain."[45] Technological advances and emerging adversarial challenges to DAF C2 functions thus necessitate a redesign of the current AOC architecture and the supporting technologies employed to meet the demands of the new digital era.

The net desired result is an evolved AOC that is capable of executing an OODA loop faster than that of the adversary and not constrained by the traditional, relatively fixed 44- to 96-hour Air Tasking Order (ATO) processing cycles. Enabled by ABMS, the AOC needs to accelerate data collection from all relevant sources, compress its processing and routing in both time and complexity, inform planning and decision-making in faster, unfixed cycles, and rapid engagement of forces to carry out the plans and bring forces to bear on the threat despite constraints imposed by force generation and logistics associated with sustained operations tempo. This must involve not only the DAF (USAF and USSF) and joint U.S. military assets, but also those of multi-national allies and partners to provide decision superiority across tactical, operational, and strategic levels of planning, command, and engagement.

---

[44] DoD, 2018, Summary of the 2018 National Defense.
[45] C.Q. Brown, Jr., 2020, *Accelerate Change or Lose*, p. 3.

ADVANCED BATTLE MANAGEMENT SYSTEM (ABMS)

What exactly is ABMS and what does it seek to do? Is ABMS a single acquisition system or a strategic concept involving multiple systems? What can it be expected to produce today and what are the challenges for realizing the full concepts behind ABMS and JADC2? The following sections provide introductory answers, supported by further details in subsequent chapters.

Evolution of ABMS

ABMS has evolved from its original inception as a C2 and surveillance system to its current construct as an enterprise-wide family of systems. As a platform, ABMS was originally introduced in 2017 as the "Airborne Battle Management and Surveillance" system, a traditional acquisition program to replace and modernize the aging Airborne Warning and Control System (AWACS) platform and a retiring fleet of E-8C Joint Surveillance and Target Attack Radar System (JSTARS).[46] In light of the NDS, however, the Air Force determined that its original plans for ABMS were no longer compatible with the objectives outlined in the NDS. DAF leaders reassessed requirements for ABMS, seeking new options for developing more robust and survivable systems that could operate within contested environments.[47] The Air Force concluded that "no single platform, such as an aircraft, would be the right solution to providing C2 capabilities across multiple domains."[48]

In April 2019, the DAF announced that the Airborne Battle Management System would become the Advanced Battle Management System—a multi-domain layered C2 family of systems (rather than a single modernization program) to strive "for the capability where any sensor can talk to any shooter whether in space, on land, at sea, in the air, or in cyberspace … [and] to perform the mission sets associated with both the JSTARS and AWACS platforms and possibly assume other roles

---

[46] See K. Osborn, 2018, "The Air Force Is Creating a System to Manage the Military's Forces in War," *The National Interest*, https://nationalinterest.org/blog/the-buzz/the-air-force-creating-system-manage-the-militarys-forces-24701, March 1, and B.W. Everstine, 2019, "USAF Selects 'Architect' for Airborne Battle Management System Program," *Air Force Magazine*, https://www.airforcemag.com/usaf-selects-architect-for-airborne-battle-management-system-program/, February 6. S.J. Freedberg, Jr., 2019, "Air Force ABMS: One Architecture to Rule Them All?" *Breaking Defense*, https://breakingdefense.com/2019/11/air-force-abms-one-architecture-to-rule-them-all, November 8.

[47] GAO (Government Accountability Office), 2020, Defense Acquisitions: Action Is Needed to Provide Clarity and Mitigate Risks of the Air Force's Planned Advanced Battle Management System, https://www.gao.gov/assets/gao-20-389.pdf, April.

[48] Ibid.

of the Theater Air Control System [and] Ground Moving Target Indicator."[49] This shift promoted a "radically new acquisition model for the Air Force" that "envisions multiple contributing programs, such as ABMS space, ABMS air, and ABMS networking and communications—each with its own funding, its own program manager, and its own schedule."[50] It also involved hiring a "Chief Architect ... to oversee the ABMS architecture design, enterprise communications and integration across programs [as well as] identify technologies to enable horizontal and vertical integration across operating environments and warfighting domains."[51]

Most recently, however, the new Secretary of the Air Force, Frank Kendall, has scrutinized the focus of ABMS. Specifically, he views ABMS as "not [having] been adequately focused on achieving and fielding specific measurable improvements in operational outcomes," and advocates instead on "developing specific, practical military technologies within a defined time."[52] He believes that the DAF needs to first determine what specific types of data and information ABMS should transmit and under which operational contexts. He also asked for performance metrics to be established to determine if ABMS is making marked improvements to current C2 capabilities, which the committee fully supports. Although the Secretary has directed a recalibration of ABMS, the committee's subsequent analysis is based on the information presented during the data gathering phase of the study conducted from late 2020 to spring 2021. So, many of the details outlined below are based on the earlier ABMS approach. However, many of the findings and recommendations remain relevant as ABMS continues on its evolutionary journey.

A Non-Traditional Acquisition Approach

As an overarching system of systems concept and visionary construct for integrating sensor-to-shooter all-domain joint command, control, communications, computers, intelligence, surveillance, reconnaissance (C4ISR) and warfighting, ABMS is composed of "a network of intelligence, surveillance, and reconnaissance

---

[49] W.B. Roper, Jr., J.M. Holmes, and D.S. Nahom, 2019, "Department of the Air Force Acquisition and Modernization Programs in the Fiscal Year 2020 National Defense Authorization President's Budget Request." Presentation to the House Armed Services Committee Subcommittee on Tactical Air and Land Forces, U.S. House of Representatives, May 2.

[50] A. McCullough, 2019, "ABMS Expected to Pick Up Speed with New Chief Architect in Place," *Air Force Magazine*, https://www.airforcemag.com/abms-expected-to-pick-up-speed-with-new-chief-architect-in-place/, March 10.

[51] W.B. Roper, Jr., J.M. Holmes, and D.S. Nahom, 2019, "Department of the Air Force Acquisition and Modernization Programs in the Fiscal Year 2020 National Defense Authorization President's Budget Request," p. 26.

[52] G. Reim, 2021, "USAF Secretary Asks 'Hard Questions' of Advanced Battle Management System," *Flight Global*, https://www.flightglobal.com/fixed-wing/usaf-secretary-asks-hard-questions-of-advanced-battle-management-system/145548.article, September 20.

sensors and will utilize cloud-based data sharing to provide warfighters with battlespace awareness for the air, land, sea, space, and cyber domains."[53] The Air Force envisions developing ABMS as an evolving "family of multiple systems"[54] and not as a single acquisition program in the traditional sense (with a set of well-defined or fixed requirements, a full cost estimate, and a single delivery schedule). Instead, ABMS seeks to leverage commercial and integrated defense capabilities—a program-of-programs or an operational concept and architecture within which individual programs will acquire specific capabilities.

Capabilities have been explored on a large scale in prior "on-ramps" or technical evaluations and demonstrations to prototype and test opportunities for leveraging commercial technologies.[55] Examples include cloud computing[56] and communication infrastructures with AI and machine learning (ML) capabilities to process and route information to commanders, decision-makers, and operators who need the information together with decision-support aids. Customized

---

[53] See CRS, 2021, Advanced Battle Management System (ABMS), https://sgp.fas.org/crs/weapons/IF11866.pdf, June 29. GAO, 2020, Defense Acquisitions: Action Is Needed to Provide Clarity and Mitigate Risks of the Air Force's Planned Advanced Battle Management System.

[54] GAO, 2020, Defense Acquisitions: Action Is Needed to Provide Clarity and Mitigate Risks of the Air Force's Planned Advanced Battle Management System.

[55] M.D. Strohmeyer, 2021, "United States Northern Command Support to ABMS," Presentation to the Air Force ABMS Committee, February 24. See also C. Pope, 2020, "Advanced Battle Management System Field Test Brings Joint Force Together Across All Domains During Second Onramp," *Air Force News*, https://www.af.mil/News/Article-Display/Article/2336618/advanced-battle-management-system-field-test-brings-joint-force-together-across/, September 3. B.W. Everstine, 2021, "USAFE's ABMS On-Ramp Included Partner Nations, Base Defense Scenario," *Air Force Magazine*, https://www.airforcemag.com/usafes-abms-on-ramp-included-partner-nations-base-defense-scenario/, March 1. "ABMS Signs More Companies Post Onramp," 2020, *Air Force News*, https://www.af.mil/News/Article-Display/Article/2359938/abms-signs-more-companies-post-onramp/, September 24. D. Henley, 2020, "Advanced Battle Management System OnRamp #2, Accelerating Data-Sharing and Decision-Making," *Defense Visual Information Distribution Service*, https://www.dvidshub.net/news/378396/advanced-battle-management-system-onramp-2-accelerating-data-sharing-and-decision-making, September 22.

[56] While the concept is evolving, the National Institute of Standards and Technology (NIST) defines cloud computing as "a model for enabling ubiquitous, convenient, on-demand network access to a shared pool of configurable computing resources (e.g., networks, servers, storage, applications, and services) that can be rapidly provisioned and released with minimal management effort or service provider interaction." See P. Mell and T. Grance, 2011, *NIST Special Publication 800-145*, "The NIST Definition of Cloud Computing," National Institute of Standards and Technology, U.S. Department of Commerce, September. Similarly, D. Mishra, Test Maintenance and Development Engineering lead at Ericcson India Private Limited, defines cloud computing as "a set of framework that provides on demand, scalable, customized, quality services in Software, platform and also provides sharable infrastructure through internet that are accessible and available everywhere." See D. Mishra, 2014, "Cloud Computing: The Era of Virtual World Opportunities and Risks Involved," *International Journal of Computer Science Engineering*, 3(4, July).

applications and hardware are being tested to interconnect sensors and systems that could not previously share information, and last mile tactical edge communications are being expanded to improve access to sensors and shooters.[57] To date, the Air Force has conducted five large-scale on-ramps to demonstrate the new C2 capabilities that it seeks to eventually field; it cancelled a sixth demonstration in March 2021 owing to budget constraints.[58]

ABMS is intended to continue leveraging evolving (primarily commercial) technologies over time rather than instantiate a static system of systems with a snapshot of extant technology. This, together with ABMS being an overarching DAF-level activity rather than a traditional C2 program, means that there is no single set of fixed requirements to build to, no single cost to estimate, and no single set of operational capabilities to field.[59] Instead, the focus has largely been on designing an enterprise-scale architecture and developing requirements to ensure that "they are met throughout the menu of systems that will comprise [ABMS]."[60] Overarching strategic requirements that lay out the JWC through JADC2 are being developed by the Joint Staff's J6 Command, Control, Communications and Computers/Cyber organization, within which specific instances of requirements, system design, and cost estimates are established for particular elements to be acquired over time.[61]

The DAF's strategic and non-traditional acquisition approach, coupled with the sizable funding requests to Congress ($136.5 million in FY 2020; $302.3 million in FY 2021; $203.8 million in FY 2022), have led to questions concerning the accounting for costs to acquire, develop, and fully integrate elements of ABMS across multiple programs and the strategy for transitioning developing technologies

---

[57] Anthologized from various presentations given by DAF representatives to the Air Force ABMS Committee from March 30 to March 31, 2021.

[58] See T. Hitchens, 2021, "Air Force Culls ABMS Experiment After Budget Cut," *Breaking Defense*, https://breakingdefense.com/2021/03/abms-hones-focus-culls-planned-experiments-in-budget-cut-wake/, March 17. V. Insinna, 2021, "Air Force curtails ABMS demos after budget slashed by Congress," *C4ISRNet*, https://www.c4isrnet.com/it-networks/2021/03/17/air-force-curtails-abms-demos-after-budget-slashed-by-congress/, March 17.

[59] P. Dunlap, 2020, Presentation to the Air Force ABMS Committee, October 30. R.G. Walden, 2021, "ABMS Perspectives from the Air Force Rapid Capabilities Office," Presentation to the Air Force ABMS Committee, January 22. See also V. Insinna, 2019, "Here's the No. 1 Rule for US Air Force's New Advanced Battle Management System," *Defense News*, https://www.defensenews.com/digital-show-dailies/paris-air-show/2019/07/09/rule-no1-for-air-forces-new-advanced-battle-management-system-we-dont-start-talking-platforms-until-the-end/, July 9.

[60] V. Insinna, 2019, "Here's the No. 1 Rule for US Air Force's New Advanced Battle Management System."

[61] S.A. Whitehead and J.S. Wellman, 2021, "Joint All Domain Command and Control," Presentation to the Air Force ABMS Committee, February 5. D.A. Crall, 2021, "Joint All Domain Command and Control," Presentation to the Air Force ABMS Committee, March 3.

into existing weapon systems.[62] The committee considers the funding requests to be appropriate based on the technical data provided by the DAF, but is concerned that the absence of clearer and more detailed program planning would challenge ABMS's ability to meet operational requirements. According to the Government Accountability Office (GAO), "weapon systems without a sound business case are at greater risk for schedule delays, cost growth, and integration issues."[63] They cited several examples of related DoD programs such as the Army's Future Combat System, the Joint Tactical Radio System, and the Transformational Satellite Communications System as evidence of cancelled programs owing to immature and under-proven technologies.[64]

Because the Air Force has not established fixed requirements nor conducted an affordability analysis for ABMS, Congress opted to slash the Air Force 2021 ABMS budget in half, allocating only $159 million of the Air Force's $302.3 million request.[65] The Secretary of the Air Force also expressed skepticism and asked for a "meaningful military capability, not just a demonstration where you show what cool thing you could do."[66] This has led to the Air Force's prioritization to shift from large-scale, on-ramp experimentations to focusing on delivering specific capabilities by allocating more than one-half of its $204 million FY 2022 budget request toward acquiring airborne datalink pods that will enable the KC-46 tanker to improve data flows between the F-35s and F-22s.[67] The DAF has also provided more details and specificity in its FY 2022 budget submission than in previous years

---

[62] Y. Tadjdeh, 2020, "Advanced Battle Management System Faces Headwinds," *National Defense Magazine*, https://www.nationaldefensemagazine.org/articles/2020/9/11/advanced-battle-management, September 11. R.S. Cohen, 2020, "Air Force Bets on ABMS in Fiscal 2021," *Air Force Magazine*, https://www.airforcemag.com/air-force-bets-on-abms-success-in-fiscal-2021/, February 11.

[63] GAO, 2020, Defense Acquisitions: Action Is Needed to Provide Clarity and Mitigate Risks of the Air Force's Planned Advanced Battle Management System.

[64] GAO, 2020, Defense Acquisitions: Action Is Needed to Provide Clarity and Mitigate Risks of the Air Force's Planned Advanced Battle Management System, p. 11.

[65] J. Keller, 2021, "Congress Cuts in Half an Air Force Battle Management System with Data Links to Join Sensors and Shooters," *Military & Aerospace Electronics*, https://www.militaryaerospace.com/communications/article/14200364/battle-management-data-links-sensors-to-shooters, March 31.

[66] V. Insinna, 2021, "New US Air Force Secretary to Shake Up Advanced Battle Management Program," *Defense News*, https://www.defensenews.com/air/2021/08/19/new-us-air-force-secretary-to-shake-up-advanced-battle-management-program/, August 19.

[67] See B.W. Everstine, 2021, "Air Force's New Plan for ABMS: Smaller Budget, Clearer Schedule," *Air Force Magazine*, https://www.airforcemag.com/air-forces-new-plan-for-abms-smaller-budget-clearer-schedule/, June 25. C. Albon, 2021, "Air Force Finalizing First ABMS Capability Release AQ Strategy, Shaping Plans for Next Release," *Inside Defense*, https://insidedefense.com/daily-news/air-force-finalizing-first-abms-capability-release-aq-strategy-shaping-plans-next-release, June 25.

and reduced their original planned budget request by more than one-half (from $449.3 million to $203.8 million).[68]

Concerns are also raised regarding the technological underpinnings of ABMS.[69] For example, what is the maturity of technology being considered, how will technology be prioritized given expected threats, uncertainties, and warfighter needs, how will legacy technologies and platforms be incorporated into newer technologies? Technology employment for its own sake does not provide value; it is how the technology addresses operational priorities against threats that matters. ABMS on-ramps (demonstrations) may be useful, but "what information will be transmitted and why, what results it is aiming to achieve and how those results will be an improvement on current command-and-control capabilities" will need to be addressed.[70]

From Demonstrations to Capabilities Releases

After nearly 2 years of on-ramp demonstrations, the former Assistant Secretary of the Air Force for Acquisition, Technology, and Logistics directed in November 2020 the transfer of ABMS from the DAF's Chief Architect's Office to the DAF's Rapid Capabilities Office (DAF RCO) as the integration program executive office (PEO).[71] In so doing, he aimed to shift the emphasis from demonstrations and

---

[68] N. Strout, 2021, "Congress Dealt ABMS a Blow, But Experts See Progress That Could Help at Budget Time," *C4ISRNet*, https://www.c4isrnet.com/battlefield-tech/c2-comms/2021/06/15/part-2-congress-dealt-abms-a-blow-but-experts-see-progress/, June 15.

[69] See GAO, 2020, Defense Acquisitions: Action Is Needed to Provide Clarity and Mitigate Risks of the Air Force's Planned Advanced Battle Management System. B. Reilly, 2021, "House Panel Praises Components of ABMS But Warns 'Questions Remain' Over Program's Direction," *Inside Defense*, https://insidedefense.com/daily-news/house-panel-praises-components-abms-warns-questions-remain-over-programs-direction, August 26. S. Sirota, 2019, "Holmes: Air Force to Accelerate ABMS Schedule to Inform FY-21, FY-22 Budget Planning," *Inside Defense*, https://insidedefense.com/daily-news/holmes-air-force-accelerate-abms-schedule-inform-fy-21-fy-22-budget-planning, June 20. S. Sirota, 2020, "Defense Spending Bill Slashes ABMS Budget Nearly in Half," *Inside Defense*, https://insidedefense.com/insider/defense-spending-bill-slashes-abms-budget-nearly-half, December 21. B. Reilly. 2021, "Air Force in 'Much Better' Place with Lawmakers Surrounding ABMS, Hinote Says," *Inside Defense*, https://insidedefense.com/insider/air-force-much-better-place-lawmakers-surrounding-abms-hinote-says, July 12.

[70] V. Insinna, 2021, "New US Air Force Secretary to Shake Up Advanced Battle Management Program."

[71] W. Roper, 2020, "Advanced Battle Management System Management Construct," Memorandum for Record, Office of the Assistant Secretary of the Air Force for Acquisition, Technology, and Logistics, https://insidedefense.com/sites/insidedefense.com/files/documents/2020/nov/11242020_abms.pdf, November 24. See also Secretary of the Air Force Public Affairs, 2020, "Air and Space Force's Acquisition Chief Appoints Rapid Capabilities Office as Integrating PEO for ABMS, Expanding from Startup Toward Rapidly Scaling Delivery Phases," *Air Force News*, https://www.af.mil/News/Article-Display/Article/2426286/air-and-space-forces-acquisition-chief-appoints-rapid-capabilities-office-as-in/, November 24.

experimentation to fielding and operationalizing ABMS capabilities. While the DAF's Chief Architect would codify ABMS technical requirements, facilitate an integrating enterprise digital architecture and standards across the DAF, establish and provide model-based systems engineering tools across the DAF, and continue directing future on-ramp demonstrations, the DAF RCO, as the chief integrating PEO, would lead in the drafting of an ABMS acquisition strategy and business case, deliver and integrate all ABMS capabilities for inclusion in architecture evaluation on-ramps, and direct the development of capability releases.[72]

In May 2021, the Air Force Chief of Staff declared that the DAF is moving to the next phase of ABMS. "Nearly two years of rigorous development and experimentation have shown beyond doubt the promise of ABMS. We've demonstrated that our ABMS efforts can collect vast amounts of data from air, land, sea, space, and cyber domains, process that information and share it in a way that allows for faster and better decisions."[73] The DAF announced the first capability release (CR1) of ABMS: fielding between 4 to 10 new datalink pods on the KC-46 Pegasus tanker to facilitate communication between the incompatible radio systems on the F-22 and the F-35 fighter jets.[74] The pods would serve as airborne hotspots connecting the two fighter jets to enable real-time communications. "The end goal isn't just 'translation' software for the fifth-generation fighters, but to continue building out the capabilities the Air Force needs to manage future All Domain Operations—from connectivity to machine-speed decision-making to real-time data sharing among commanders in far flung [headquarters] HQs."[75] Furthermore, the Air Force Chief of Staff seeks not only to push information to tactical edge command centers, but also to bring data back. "Each one of our platforms has some level of data on it … but sometimes, it's tied to that platform and doesn't get off the platform until you get it back on the ground. Why wait several hours to get it back … when you can

---

[72] Secretary of the Air Force Public Affairs, 2020, "Air and Space Force's Acquisition Chief Appoints Rapid Capabilities Office as Integrating PEO for ABMS, Expanding from Startup Toward Rapidly Scaling Delivery Phases," *Air Force News*, https://www.af.mil/News/Article-Display/Article/2426286/air-and-space-forces-acquisition-chief-appoints-rapid-capabilities-office-as-in/, November 24.

[73] C. Pope, 2021, "With Its Promise and Performance Confirmed, ABMS Moves to a New Phase," *Air Force News*, https://www.af.mil/News/Article-Display/Article/2627008/with-its-promise-and-performance-confirmed-abms-moves-to-a-new-phase/, May 21.

[74] N. Miknev, 2021, "ABMS Capability Release 1," Presentation to the Air Force ABMS Committee, March 30. See also T. Hitchens, 2021, "First ABMS Buy: KC-46 Pods to Link F-22, F-35," *Breaking Defense*, https://breakingdefense.com/2021/06/first-abms-buy-kc-46-pods-to-link-f-22-f-35/, June 25, and A. McCullough, 2021, "ABMS, in New Phase, Prepares to Start Fielding," *Air Force Magazine*, https://www.airforcemag.com/abms-in-new-phase-prepares-to-start-fielding/, May 21.

[75] T. Hitchens, 2021, "First ABMS Buy: KC-46 Pods to Link F-22, F-35."

actually push that information and data around real time to drive decisions?"[76] The DAF plans to invest $170 million this fiscal year to execute CR1.[77]

While not official, capability release two (CR2) will likely "use cloud-computing, fiber-optic networks, AI, and other new technologies" to accelerate homeland defense missions and decision-making in support of U.S. Northern Command (USNORTHCOM) and the North American Aerospace Defense Command (NORAD).[78] Many of these largely commercial capabilities were tested and demonstrated during the large-scale demonstration, ABMS on-ramp 2, conducted in fall 2020. The exercise, known as Shadow Operations Center-Nellis (ShOC-N), established a virtual environment for commercial vendors and DoD participants to operate in and connect with homeland defense agencies to provide a common operating picture (COP). Vendors tested their connectivity and ability to provide real-time situation awareness within the USNORTHCOM battlespace.[79]

Moving forward, the DAF RCO anticipates building on successful capability demonstrations in future on-ramps and introducing new digital capabilities in upcoming capability releases. According to its director, "To build ABMS, you must first build the digital structures and pathways over which critical data is stored, computed, and moved. The DAF needs a smart, fast, and resilient 'system of systems' to establish information and decision superiority, and ABMS will be that solution."[80]

ABMS as a Contributor to JADC2

As the DAF's contribution to JADC2, ABMS is designed to be "an ecosystem of sensors, fusion, and data-transfer networks aided by cloud-based processing power and AI that will empower modern C2."[81] The goal is to enable disjointed

---

[76] M. Jasper, 2021, "The Advanced Battle Management System Is Ready for Real-World Testing, the Service Announced," *NextGov*, https://www.nextgov.com/emerging-tech/2021/05/air-forces-jadc2-contribution-shifting-operational-status/174291/, May 25.

[77] N. Miknev, 2021, "ABMS Capability Release 1," Presentation to the Air Force ABMS Committee, March 30.

[78] B.W. Everstine, 2021, "Air Force's New Plan for ABMS: Smaller Budget, Clearer Schedule," *Air Force Magazine*, https://www.airforcemag.com/air-forces-new-plan-for-abms-smaller-budget-clearer-schedule/, June 25.

[79] M.D. Strohmeyer, 2021, "United States Northern Command Support to ABMS," Presentation to the Air Force ABMS Committee, February 24.

[80] A. McCullough, 2021, "ABMS, in New Phase, Prepares to Start Fielding." R.G. Walden, 2021, "ABMS Perspectives from the Air Force Rapid Capabilities Office," Presentation to the Air Force ABMS Committee, January 22.

[81] D.A. Birkey, 2021, "The Battle for the Soul of JADC2," *Air Force Magazine*, https://www.airforcemag.com/article/the-battle-for-the-soul-of-jadc2/, April 23.

and often incompatible equipment to seamlessly and securely communicate across all domains.

As a technical solution, ABMS provides a connection and understanding across all battlespaces and domains without consideration to seams to ensure that "all 11 combatant commands, can be operating off of this same level of understanding. They can plan together and then they can execute together."[82] Technical capabilities are advanced via a DevSecOps rapid development approach that refreshes every 4 months. According to DAF leaders, ABMS will enable JADC2 by "simultaneously sensing, making sense of and acting upon a vast array of data and information from [all] domains, fusing and analyzing the data with the help of ML and AI and providing warfighters with preferred options at speeds not seen before."[83]

While DoD strives for network connectivity across all domains, much of the near-term focus has been on establishing training and doctrine related to JADC2. At the 2021 DAF Command and Control Summit led by the Commander of Air Combat Command (ACC), discussions centered on "the need to look at how to leverage advanced technology and AI through innovations in doctrine and training that optimize the speed of decision-making, organizational structures scaled to leverage technological innovation and efficiencies toward winning across the spectrum of competition and conflict, and continuing to develop the all-domain skills and decision-focused leaders needed to plan and execute JADC2."[84] To address these requirements, the Air Force's cross-functional lead for joint warfighting integration recently announced the completion of a *JADC2 Supporting Concept* that guides the DAF's "concept-driven, threat-informed JADC2 capability development to include doctrine, training materiel, and personnel."[85] Furthermore, the Air Force established a new 13O Air Force Specialty Code (AFSC) designed to secure continued dominance in the air, space, and cyberspace domains. Individuals who are coded as 13O are trained to plan and execute multi-domain operations at the operational level across multiple warfighting domains. The goal of these collective efforts is to "build our people into informed, decisive leaders who can plan and

---

[82] J. Eddins, 2021, "Valenzia: ABMS Will Deliver the 'Decision Advantage,'" *Airman Magazine*, https://www.macdill.af.mil/News/Features/Display/Article/2647112/valenzia-abms-will-deliver-the-decision-advantage/, May 26.

[83] C. Pope, 2021, "With Its Promise and Performance Confirmed, ABMS Moves to a New Phase," *Air Force News*, https://www.af.mil/News/Article-Display/Article/2627008/with-its-promise-and-performance-confirmed-abms-moves-to-a-new-phase/, May 21.

[84] N.E. Mathison, 2021, "2021 C2 Summit Enhances Air Force Contribution to Joint All-Domain Command and Control," *Air Force News*, https://www.af.mil/News/Article-Display/Article/2476713/2021-c2-summit-enhances-air-force-contribution-to-joint-all-domain-command-and/, January 20.

[85] J. Barnett, 2021, "Air Force Inks New ABMS Concept Document," *FedScoop*, https://www.fedscoop.com/air-force-abms-concept-document-signed-jadc2/, July 21.

execute in a joint, high-tech environment where AI and ML are also contributing to the fight alongside them."[86]

To further direct ABMS and its support to JADC2, the Chief of Staff of the U.S. Air Force, Gen. Charles Q. Brown, Jr., signed the *ABMS Campaign Plan* in May 2021. The plan includes eight warfighting capabilities that the DAF seeks to accomplish to achieve decision superiority: (1) data sharing; (2) human capital development; (3) distributed decision-making; (4) advanced communications; (5) advanced sensing; (6) integrated planning; (7) C2 of convergence of effects; and (8) accelerated decision-making.[87] Together, these capabilities would enable ABMS to securely collect and transmit volumes of data from air, land, sea, space, and cyber domains, as well as process information and share it amongst the Joint Force and multi-national partners—the cornerstone of the JADC2 mission.

Other Contributors to JADC2 and Complicating Factors

Originally advanced by DAF leaders as the chief JADC2 solution for the DoD, other military Services and DoD agencies have since proposed their own inputs to JADC2. Both the Army and the U.S. Department of the Navy (DoN) have launched similar efforts that parallel ABMS. Each seeks to prototype and experiment with technologies and operational approaches to support the all-domain JWC. The Army is enabling joint and combined overmatch and addressing the demands of the Joint Operating Environment through Project Convergence (PC),[88] and the DoN is pursuing Project Overmatch to develop the network, infrastructure, data

---

[86] N.E. Mathison, 2021, "2021 C2 Summit Enhances Air Force Contribution to Joint All-Domain Command and Control."

[87] J. Valenzia, 2021, "The Ability to Share Data Could Prove Key to Deterring and Defeating Adversaries," *C4ISRNet*, https://www.c4isrnet.com/opinion/2021/05/29/the-ability-to-share-data-could-prove-key-to-deterring-and-defeating-adversaries/, May 29.

[88] For more on Project Convergence, see A. Abadie, 2021, "Project Convergence Overview," Presentation to the Air Force ABMS Committee, January 8, and Army Futures Command, "Project Convergence," https://armyfuturescommand.com/convergence/. See also J. Lacdan, 2021, "Project Convergence 21 to Showcase Abilities of the Joint Force," *Army News Service*, https://www.army.mil/article/249422/project_convergence_21_to_showcase_abilities_of_the_joint_force, August 15. T. South, 2021, "New in 2021: The Army's Project Convergence Scales Up," *Army Times*, https://www.armytimes.com/news/your-army/2021/01/04/new-in-2021-the-armys-project-convergence-scales-up/, January 4. J. Judson, 2020, "Inside Project Convergence: How the US Army Is Preparing for War in the Next Decade," *Defense News*, https://www.defensenews.com/smr/defense-news-conference/2020/09/10/army-conducting-digital-louisiana-maneuvers-in-arizona-desert/, September 10.

architecture, tool, and analytics to support maritime dominance and interoperability in JADC2.[89]

While the objective of supporting JADC2 is central to each development, each Service is adopting considerably different approaches. For example, PC is proposed as "a campaign of learning to aggressively pursue an AI and ML-enabled battlefield management system"[90] and is designed around five core elements: people, weapons systems, C2, information and terrain.[91] PC emphasizes building lethality at scale by leveraging a mix of AI, robotics, and autonomy.[92] The organization responsible for leading the effort, the Army Futures Command (AFC), plans to run PC on an annual cycle "achieving objectives from frequent experiments with technology, equipment, and solder feedback throughout the year and culminating in an annual exercise or demonstration."[93]

In contrast, the Navy "envisions a future fleet with manned and unmanned ships, submarines and aircraft operating in a dispersed manner and collecting a ton of data to fill in a COP—which operational commanders could then use to, if ever needed, have the best sensor platform send targeting data to the best shooter to attack an enemy."[94] Project Overmatch is the "Navy's effort to create a 'Naval Operational Architecture' to link ships to Army and Air Force assets,"[95] and employs an engineering development approach to "enable a Navy that swarms the sea, delivering synchronized lethal and nonlethal effects from near-and-far, every

---

[89] For more on Project Overmatch, see D.W. Small, 2021, "Project Overmatch," Presentation to the Air Force ABMS Committee, March 3. M. Shelbourne, 2020, "Navy's 'Project Overmatch' Structure Aims to Accelerate Creating Naval Battle Network," *USNI News*, https://news.usni.org/2020/10/29/navys-project-overmatch-structure-aims-to-accelerate-creating-naval-battle-network, October 29. J. Barnett, 2021, "Top Navy Officer Says Project Overmatch Work 'Headed in the Right Direction,'" *FedScoop*, https://www.fedscoop.com/top-naval-officer-not-satisfied-with-progress-on-project-overmatch/, August 2. A. Eversden and D. Larter, 2021, "Exclusive: Navy Transfers Network Authorities to Project Overmatch Office," *C4ISRNet*, https://www.c4isrnet.com/battlefield-tech/it-networks/2021/03/05/exclusive-navy-transfers-network-authorities-to-project-overmatch-office/, March 4. L.C. Williams, 2021, "Navy Aims to Tackle Cross-Domain Data Sharing in Project Overmatch," *FCW*, https://fcw.com/articles/2021/08/03/sas-overmatch-data-sharing-navy.aspx, August 3.

[90] Army Futures Command Project Convergence website, https://armyfuturescommand.com/convergence/, accessed August 6, 2021.

[91] A. Abadie, 2021, "Project Convergence Overview," Presentation to the Air Force ABMS Committee, January 8.

[92] T. South, 2021, "New in 2021: The Army's Project Convergence Scales Up," *Army Times*, January 4.

[93] CRS (Congressional Research Service), 2020, *The Army's Project Convergence*, https://sgp.fas.org/crs/weapons/IF11654.pdf, October 8.

[94] M. Eckstein and M. Shelbourne, 2021, "Navy to Field Early 'Project Overmatch' Battle Network on Theodore Roosevelt CSG in 2023," *USNI News*, https://news.usni.org/2021/02/08/navy-to-field-early-project-overmatch-battle-network-on-theodore-roosevelt-csg-in-2023, February 10.

[95] CRS, 2021, Joint All-Domain Command and Control: Background and Issues for Congress.

axis, and every domain."⁹⁶ The goal is to "develop the networks, infrastructure, data architecture, tools, and analytics that support the operational and developmental environment that will enable our sustained maritime dominance."⁹⁷ According to the Commander, Naval Information Warfare Systems Command and the Director of Project Overmatch, the DoN will leverage and integrate the latest in digital technologies to include AI, ML, and information and networking technologies into existing DoN networks and platforms for achieving improved global fleet readiness.⁹⁸ This goal is not necessarily to solely acquire new solutions, but to capitalize on improving the operational effectiveness of existing C2 networks and platforms.

Beyond the military Services, the Defense Advanced Research Projects Agency (DARPA) has established Mosaic Warfare⁹⁹ and the Office of the Secretary of Defense (Research and Engineering) has established fully networked command, control, and communications (FNC3), as their contributions to JADC2.¹⁰⁰ The U.S. Special Operations Command (SOCOM) is also developing its own Special Operations Forces (SOF)-specific data management environment (data fabric) with a set of common standards and tools that will enable SOF systems to communicate with one another.¹⁰¹ Each seeks to enable key commercial technologies to improve and enhance C2 operations; but again, little to no coordination is being executed to ensure an enterprise-wide solution.

The challenge, of course, is that each of these efforts is experimenting with joint interoperability involving capabilities and assets outside of their respective services and agencies, so eventual control and jurisdictional questions will arise. Moreover, while all of these efforts are loosely coordinating, there has been little to no reconciliation to understand exactly whose approach will apply, using which system, operating in which battlespace domain, and projecting against which adversary threatening which Joint and Service C2 posture. This challenge is further complicated in that JADC2 is designed to involve multi-national allied partners in planning (rather than as an afterthought once fielded), but those considerations

---

⁹⁶ M. Gilday, 2020, "Memorandum to Rear Admiral Douglas W. Small, United States Navy on Project Overmatch," U.S. Department of the Navy, https://insidedefense.com/sites/insidedefense.com/files/documents/2020/oct/10192020_overmatch.pdf, October 1.
⁹⁷ Ibid.
⁹⁸ D.W. Small, "Project Overmatch," 2021, Presentation to the Air Force ABMS Committee, March 3.
⁹⁹ For more on Mosaic Warfare, see "DARPA Tiles Together a Vision of Mosaic Warfare," https://www.darpa.mil/work-with-us/darpa-tiles-together-a-vision-of-mosiac-warfare.
¹⁰⁰ CRS, 2021, Joint All-Domain Command and Control: Background and Issues for Congress.
¹⁰¹ A. Eversden, 2021, "SOCOM Data Official: Build Interoperability into New Systems for Joint War Fighting," *C4ISRNet*, https://www.c4isrnet.com/battlefield-tech/it-networks/2021/05/26/socom-data-official-build-interoperability-into-new-systems-for-joint-war-fighting/, May 26. See also S. Magnuson, 2021, "SOFIC NEWS: SOCOM Looking to Break Barriers to Deliver Data Globally," *National Defense Magazine*, https://www.nationaldefensemagazine.org/articles/2021/5/20/socom-looking-to-break-barriers-to-deliver-data-globally, May 20.

are largely nascent compared to the unresolved differences between the concepts and systems from the military Services and supporting agencies.[102] Accordingly, without proper coordination, a clear delineation of roles and responsibilities, and common operating standards, the risk is that "each service, COCOM [combatant command], or agency goes in its own direction and develops multiple stove-piped networks that do not allow the kind of interoperability and resilience that would be possible with a more coordinated approach."[103]

In the following chapters, the committee examines in detail planned ABMS data and communications architecture, reviews the proposed governance approach and supporting processes and recommends a path to address the identified technical gaps and process improvements to achieve ABMS capabilities more effectively. The final chapter will summarize the committee's major findings and recommendations.

---

[102] See J. Garamone, 2021, "Joint All-Domain Command, Control Framework Belongs to Warfighters," *DoD News*, https://www.defense.gov/Explore/News/Article/Article/2427998/joint-all-domain-command-control-framework-belongs-to-warfighters/, November 30. G.I. Seffers, 2020, "Army Suggests Adding Five Eyes Nation Allies in JADC2," *SIGNAL*, https://www.afcea.org/content/army-suggests-adding-five-eyes-nation-allies-jadc2, July 14.

[103] T. Harrison, 2021, "Battle Networks and the Future Force," Center for Strategic and International Studies, https://www.csis.org/analysis/battle-networks-and-future-force, August 5.

2

Architecture and Data

The capabilities that we're building and using, we're actually designing into it the capability to snap together like LEGO blocks, both our Air Force capabilities, as well as our sister services and international partners.... The power of this architecture is unlocked by services, allies and partners working together to connect networks and share information at machine speed. That's all-domain superiority.
 —Preston Dunlap, Chief Architect, U.S. Department of the Air Force[1]

ARCHITECTURE OVERVIEW

The architecture of the Advanced Battle Management System (ABMS) refers to the relationships and interconnections between individual system components and capabilities. The granularity of the architecture is typically described at the system and sub-system level, with multiple capabilities represented within each sub-system component. These sub-components are modularized and interconnected through a variety of technologies, including wired and wireless communications, interconnect frameworks within a single platform, satellite communication (SATCOM) systems, and commercial telecommunications. The intent is to integrate all components and

---

[1] S. Freedberg, Jr., 2019, "Air Force ABMS: One Architecture to Rule Them All?" *Breaking Defense*, https://breakingdefense.com/2019/11/air-force-abms-one-architecture-to-rule-them-all, November 8, and J. Lacdan, 2020, "Army, Air Force Form Partnership, Lay Foundation for CJADC2 Interoperability," *Army News Service*, October 1.

interconnections to achieve the larger system objectives—in this case, the coordinated command and control (C2) across all Department of the Air Force (DAF) sensors, network components, and weapon systems, as well as connections to the larger Joint All-Domain Command and Control (JADC2) enterprise architecture. Figure 2.1 provides an overview of ABMS and Figure 2.2 details the ABMS concept.

As a C2 family of systems, ABMS involves both data processing and communication that support computation, sensing, and actuation, defined as the application of weapons effects.[2] Under the current construct, each C2 node has the ability to provide autonomous computation in the use of sensors, data, and actuation. The success of this architecture fundamentally depends on the rigorous adherence to Application Programming Interface (API) and data standards that provide a common application environment and a set of flexible protocols.

ABMS architecture is also intended to closely link to elements of JADC2, although the committee saw little evidence of this. The focus appears to be largely on defining ABMS-specific platforms, components, and interconnections that are native to the DAF. The committee strongly encourages developing ABMS architecture at the Joint level to achieve interoperability with other Services and multinational partners.

UNCLASSIFIED

Advanced Battle Management System
Program Overview

- Create secure military digital network environment leveraging proven digital infrastructure, commercial technologies, and applications
  - Define distributed cloud, network management, global footprint for military applications
- Connect the joint force to enable all domain dynamic operations
  - Build the digital infrastructure that connects the Joint Warfighting force
  - Enable sharing of information across USAF, USSF, Joint, and multi-domains
  - Will provide situational awareness to enable better operational decisions as well as tactical decision making and execution at speed
- ABMS acquisition efforts focus on six capabilities:
  1. *Secure Processing
  2. *Connectivity
  3. *Data Management
  4. Applications
  5. Sensor Integration
  6. Effects Integration

* Digital Infrastructure

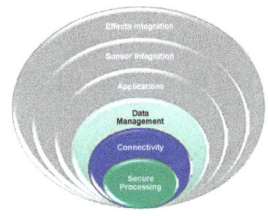

UNCLASSIFIED

FIGURE 2.1 Advanced Battle Management Program Overview. SOURCE: Department of the Air Force's Rapid Capabilities Office. Approved for public release.

[2] R. Walden, 2021, "ABMS Perspectives from the Air Force Rapid Capabilities Office," Presentation to the Air Force ABMS Committee, January 22.

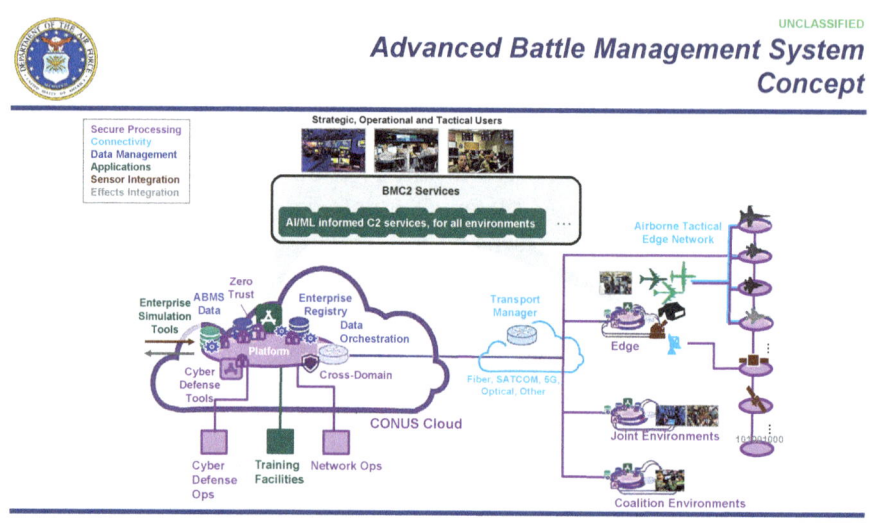

FIGURE 2.2 Advanced Battle Management System Concept. SOURCE: Department of the Air Force's Rapid Capabilities Office. Approved for public release.

A standard that is open, modular, and (most importantly) scalable also needs to be defined to ensure that Army, Navy, Marine Corps, DoD agencies, and multinational partner assets (or their portion of the JADC2 system) can effectively integrate with the ABMS architecture as envisioned by the JWC. Innovative and "smart" systems will also need to integrate and/or interoperate with legacy systems. Additionally, because the majority of U.S. weapons systems are designed and supplied by prime system integrators in the defense industrial base, they should be engaged throughout the development process.

FINDING 1: To support the JWC, the ABMS architecture must be considered integral to JADC2, which needs to be clearly defined.

FINDING 2: An open, modular design is needed to support evolution of ABMS and JADC2.

RECOMMENDATION 1: The Department of the Air Force Chief Architect's Office and the Department of the Air Force Rapid Capabilities Office should define the Advanced Battle Management System (ABMS) architecture at the Joint All-Domain Command and Control level to ensure interoperability with other ABMS-like systems being developed.

RECOMMENDATION 2: The Joint Staff J6 or a designated U.S. Department of Defense executive agent should establish interoperability requirements and performance metrics for all participants in Joint All-Domain Command and Control to allow for eventual integration of all capabilities.

RECOMMENDATION 3: The Department of the Air Force Chief Architect's Office and the Department of the Air Force Rapid Capabilities Office should design the Advanced Battle Management System architecture to be modular and include open standards and interfaces that would enable configuration with other Service variants.

The ABMS architecture should include the ability to achieve integrity, availability, and confidentiality for all communications, data, and computation across all applications. The goal is to establish not only a common operating picture (COP), but a representation of the totality of data that could be used for any strategic or tactical decision at the edge. If the tactical edge is sufficiently served with relevant data, velocity of commanders' intent can be maintained at some level. If the tactical edge is not serviced, but the strategic level is, relevant actions and its resultant effects will be limited at best, and ineffective at worst.

Emphasis should also be given to communications, data, and computation operating in degraded or denied environments, where a level of autonomy for disconnected or high-latency components need to operate uninterrupted. More importantly, protecting ABMS—both systems and data—against cyber vulnerabilities and adversarial attacks requires that cybersecurity be included as part of the overall architecture design.

FINDING 3: The ABMS architecture requires integrity, availability, and confidentiality for all communications, data, and computation elements.

RECOMMENDATION 4: The Department of the Air Force Chief Architect's Office and the Department of the Air Force Rapid Capabilities Office should design the Advanced Battle Management System's architecture with specific technical requirements and solutions for ensuring that communications, data, and computation may continue to operate in degraded or denied access environments.

Architecture and Technology Status

In rudimentary terms, the ABMS architecture consists of an array of platforms, sensors, networks, and datalinks interconnected through a secure cloud[3] to facilitate sensing, sense-making, and acting within the Joint All-Domain context. Figure 2.3 depicts the ABMS architecture.

ABMS is a family of systems that includes both hardware and software supported by technologies to input and connect with the JADC2 network. Under the original governance by the DAF's Chief Architect, it comprises six product (i.e., technologies and capabilities) categories:[4]

1. *Sensor integration*, including sensors on satellites and aircraft, ground-based radar, among others;
2. *Data and data management*;
3. *Secure processing* that involves both cybersecurity and the ability to transmit and process data across all classifications while allowing broad access to data products;

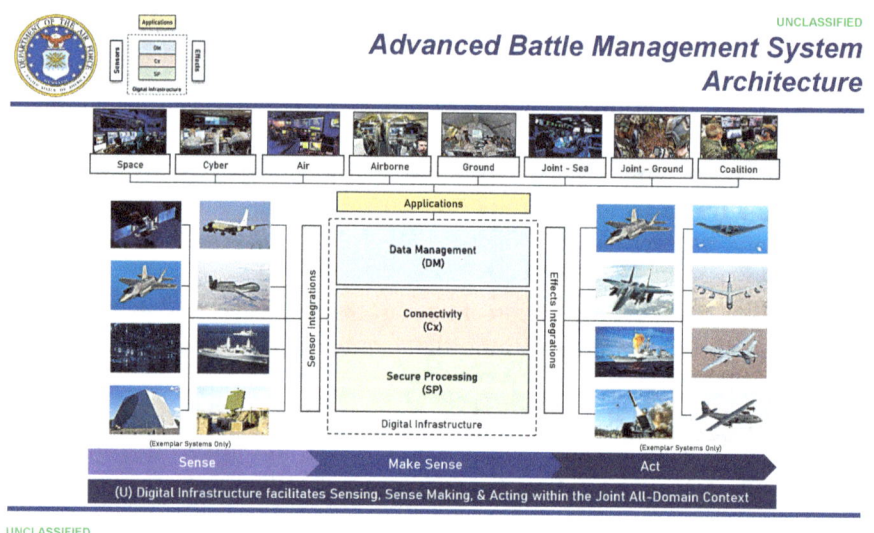

FIGURE 2.3 Advanced Battle Management System architecture. SOURCE: Department of the Air Force's Rapid Capabilities Office. Approved for public release.

---

[3] For more on the Air Force cloud, see Assistant Secretary for Acquisition, Technology, and Logistics, 2019, "Memorandum for Acquisition Workforce: Cloud One and DevSecOps," September 13.

[4] D. Mayer, 2021, "ABMS Aims to Revolutionize Data Flow, Speed Decisions," *Air Force News*, https://www.af.mil/News/Article-Display/Article/2559022/abms-aims-to-revolutionize-data-flow-speed-decisions/, April 1.

4. *Connectivity*, from both military and commercial networks, including artificial intelligence (AI) and machine-to-machine links across both legacy and new weapons platforms;
5. *Applications*; and
6. *Effects integration.*

Each of these original product categories or capabilities was developed through iterative 4-month "demonstration sprints" conducted during on-ramp experiments.[5] At the time of presentation by the DAF's Chief Architect to the committee, they are supported by multiple software product lines to include:

- cloudONE, the secure cloud that supports multi-level classifications across the ABMS enterprise;
- edgeONE, a local cloud backup in the event that datalinks are disconnected from cloudONE;
- dataONE, a database that builds on the Unified Data Library (UDL);[6]
- crossDomainONE, which enables the transmission of data across classification levels;
- omniaONE, a common operating picture of the all-domain battlefield using multiple feeds merged through a system called fuseONE, which is a cloud-based fusion environment;
- AI/smartONE, which layers on top of omniaONE and uses artificial intelligence to cue the user to potentially useful information;
- feedONE, cloud-based data feeds from all sources;
- commandONE, a battle management command and control system using the Link16e network;[7] and
- gatewayONE, a communications gateway designed to secure two-way data path across multiple platforms and domains.[8]

---

[5] W. Roper, 2020, "Advanced Battle Management System Management Construct."

[6] UDL is "a collection of space objects that integrates data from commercial and government sources." See S. Erwin, 2021, "Bluestaq Wins $280 Million Space Force Contract to Expand Space Data Catalog," *Space News*, https://spacenews.com/bluestaq-wins-280-million-space-force-contract-to-expand-space-data-catalog/, March 23. See also F. Wolfe, 2021, "Unified Data Library to Be Significant Part of U.S. Space Force Contribution to ABMS," *Defense Daily*, https://www.defensedaily.com/unified-data-library-significant-part-u-s-space-force-contribution-abms/space/, May 24.

[7] Link 16 is "the DOD's primary tactical data link for all military service and defense agency command, control and intelligence systems." See B.E. White, 1999, "Tactical Data Links, Air Traffic Management, and Software Programmable Radios," *Proceedings of the IEEE*, https://www.mitre.org/sites/default/files/publications/white_tactical_data_links.pdf. Linked16e is the enhanced Link 16.

[8] P. Dunlap, 2020, "ABMS Overview," Presentation to the Air Force ABMS Committee, October 30.

ABMS is also supported by a variety of hardware product lines to include:

- radioONE, a new radio frequency antenna for receiving SATCOM data;
- apertureONE, a common aperture for communications and radar;
- boxONE, the workstation used to access cloudONE or edgeONE;
- phoneONE, a smartphone that accesses cloudONE or edgeONE; and others.[9]

The goal is to connect all of these elements into a holistic data-integration and command decision enabler to accelerate the find, fix, target, track, engage, and assess (FFTTEA) kill chain. The committee makes no determination regarding each of these product categories, because it was not provided with information regarding their overall performance.

The ABMS architecture remains nascent and continuously evolving. It is difficult to comprehend the totality of the ABMS architecture without examining each component, which is being constructed from the bottom-up. "When we say 'build bottom-up,' we mean that this process of iterative experimentation should occur at the tactical level. The end users of ABMS—joint sensors and shooters—should be the ones cycling through new techniques and technologies that will come to form ABMS."[10] It is important to note that while each component is developed as a building block, ABMS is designed to be a vast ecosystem that integrates each block into its core architecture. As new technologies and capabilities emerge and are experimented in on-ramp exercises, proven technologies will be fielded in capability releases. As such, the ABMS architecture is evolving and in the early stages of definition.

As a possible connective framework, a time-triggered architecture (TTA) may be a viable option.[11] TTA is an integration framework that provides an environment for integrating components into a system, where certain properties of the system are guaranteed for the system by the framework independently of the components.[12]

---

[9] Ibid.

[10] P. Birch, R. Reeves, and B. DeWees, 2020, "Build ABMS from Bottom-Up, for the Joint Force," *Breaking Defense*, https://breakingdefense.com/2020/05/build-abms-from-bottom-up-for-the-joint-force/, May 13.

[11] J. Valenzia, 2020, "JADC2 and ABMS," Presentation to the Air Force ABMS Committee, December 18. For more on TTA, see H. Kopetz, 1997, "Why a Distributed Solution?," p. 34, in *Real-Time Systems: Design Principles for Distributed Embedded Applications*, Springer. See also H. Kopetz and G. Bauer, 2002, "The Time-Triggered Architecture," *Proceedings of the IEEE Special Issue on Modeling and Design of Embedded Software*, October.

[12] J. Rushby, undated, "An Overview of the Time Triggered Architecture (TTA) and Its Formal Verification," http://www.csl.sri.com/users/rushby/slides/kestrel05.pdf, Computer Science Laboratory, SRI International.

Within this framework is the capability to rapidly adapt to changing conditions, mission requirements, technology improvements, and threats.[13] The advantages of using TTA are "to precisely specify the interfaces among the nodes, to simplify the communication and agreement protocols, to perform prompt error detection, and to guarantee the timeliness of real-time applications."[14] It has been used in automobile applications (Audi, Peugeot S.A., and others), but also for aircraft applications (Honeywell Aerospace).[15]

Furthermore, to support the aims of JADC2, the ABMS architecture needs to remain evolvable, allowing for continuous development, deployment, testing, refinement, and improvement over time. While DAF leaders have embraced flexibility in developing ABMS, Congress's decision to reduce the ABMS FY 2021 budget by nearly one-half may hamper the department's ability to continue with this approach.[16] According to the Acting Assistant Secretary of the Air Force for Acquisition, Technology, and Logistics, "We're doing everything in our power to keep efforts going ... with a cut of that magnitude, it will have an impact."[17]

The ABMS architecture must focus on protecting data, computations, and communications from adversary access or manipulation; providing high-bandwidth communications with significant redundancy and resiliency; deploying advanced computational platforms with surplus or expanding capacity to handle thousands of parallel C2 tasks simultaneously; providing large data curation and storage for near-real-time data accessibility to all decision-support nodes; and allowing for interoperability with both legacy platforms and inter-Service systems and networks. Furthermore, the architecture should remain modular with standardized, open interfaces based on practicality, and ideally, a track record of successful implementation, so as technology progresses, new components may be easily incorporated into ABMS to improve realization of all mission requirements. The use of open standards and APIs would facilitate ready and rapid adoption of new technologies that are critical for enabling the integration of future communications, computation, data, and software improvements.

FINDING 4: The ABMS architecture must be adaptive to enable continuous development, deployment, testing, refinement, and improvement over time.

---

[13] E. Bryant, 2021, "Cybersecurity in JADC2 and Contested Environments," Presentation to the Air Force ABMS Committee, April 16.

[14] H. Kopetz and G. Bauer, 2002, "The Time-Triggered Architecture," p. 1.

[15] Rushby, TTA Overview 4.

[16] See T. Hitchens, 2021, "Air Force Working to Minimize Damage from ABMS Budget Cut," *Breaking Defense*, https://breakingdefense.com/2021/02/air-force-working-to-minimize-damage-from-abms-budget-cut/, February 24.

[17] Ibid.

FINDING 5: Under current plans, the ABMS architecture is likely to face interoperability challenges if it is to fully realize the JADC2 JWC.

Technology for Data-Centric Operations

Current DoD communications are primarily point-to-point.[18] Advancing the command, control, communications, computers, intelligence, surveillance, and reconnaissance (C4ISR) architecture across a complex and dispersed DoD enterprise to take advantage of newer, more agile networked communication solutions at both strategic and tactical levels will require a significant amount of time, resources, and commitment. As increasing numbers of operations shift to data-centricity, it is critical that data—more specifically, the access, storage, transmission, validation, and protection of data—be considered in the enterprise architecture design.

There are an increasing number of modern tools to support data-centric operations. These range from tools to support application-to-application interfaces to the use of clouds to provide on-demand computing, storage, and sharing.[19] Others include predictive analytics, data virtualization, stream analytics, distributed storage, data preprocessing, and others.[20] While the work on capabilities such as DARPA's data translation and rapid software integration tool known as system of systems technology integration tool chain for heterogeneous electronic systems (STITCHES) could support architectural transitions and facilitate tactical operations, ABMS must also support data and computing needs for highly sensitive strategic missions to include nuclear command, control, and communications (NC3) operations. Adopting an array of data-exchange technologies that could support the spectrum of capabilities should remain a central objective of ABMS's architecture design.

Furthermore, as AI/machine learning (ML) technologies become ever more pervasive and capable, machine-to-machine data sharing—at scale within accept-

---

[18] J.C. Stenbit, 2021, "DoD C3I Perspectives," Presentation to the Air Force ABMS Committee, February 24.

[19] A good example of application-to-application interfaces is DARPA's System of Systems Technology Integration Tool Chain for Heterogeneous Electronic Systems (STICHES). See DARPA (Defense Advanced Research Projects Agency), 2020, "Creating Cross-Domain Kill Webs in Real Time," Defense Advanced Research Project Agency website, https://www.darpa.mil/news-events/2020-09-18a, September 18.

[20] See Maruti Techlabs, 2017, "10 Key Technologies That Enable Big Data Analytics for Businesses," *Toward Data Science*, https://towardsdatascience.com/10-key-technologies-that-enable-big-data-analytics-for-businesses-d82703891e2f, September 26, and R. Sheldon, 2021, "Why and How to Adopt a Data-Centric Architecture," *TechTarget*, https://searchconvergedinfrastructure.techtarget.com/tip/Why-and-how-to-adopt-a-data-centric-architecture, January 28.

able timelines for both planned and unanticipated mission requirements—will become increasingly important and must be fully integrated into the ABMS and JADC2 architecture design. Present-day machine-to-machine data sharing is primarily point-to-point and utilized in support of planned and anticipated missions with predetermined data feeds. However, as adversaries rapidly advance their own use of AI/ML, the DoD and the DAF need to implement advanced AI capabilities that would improve C2 and time-sensitive decision-making from the range of tactical operations to the strategic planning level.[21] Implementing a robust machine-to-machine data sharing capability will require, at a minimum:

- Appropriate placement of data storage and computing to facilitate timely access to—and availability of—data to support military operations;
- Security and synchronization of data storage; and
- Redundancy to support operations in degraded environments and to enable reconstitution; among others.

RECOMMENDATION 5: The Department of the Air Force Rapid Capabilities Office should adopt an array of data-exchange technologies that could support the entire spectrum of capabilities, from tactical to strategic.

Highly Capable Processing: AI and ML

Communications must not only be networked, but also enable direct connections and relays—both with and between—humans and machines. As a C2 network, ABMS requires rapid, accurate, secure, and resilient data processing. This will involve a large volume of complex data and event processing and will increase the time sensitivity and demand for quality of services, particularly in contested areas with jamming and poor communications. Without the aid of machines, process-

---

[21] For examples, see Y. Tadjdeh, 2021, "Algorithmic Warfare: Russia Expanding Fleet of AI-Enabled Weapons," *National Defense Magazine*, https://www.nationaldefensemagazine.org/articles/2021/7/20/russia-expanding-fleet-of-ai-enabled-weapons, July 20, and A. Eversden, 2021, "A Warning to DoD: Russia Advances Quicker Than Expected on AI, Battlefield Tech," *C4ISRNet*, https://www.c4isrnet.com/artificial-intelligence/2021/05/24/a-warning-to-dod-russia-advances-quicker-than-expected-on-ai-battlefield-tech/, May 24. See also S. Bendett, 2019, "Russia's AI Quest Is State-Driven—Even More Than China's. Can It Work?" *DefenseOne*, https://www.defenseone.com/ideas/2019/11/russias-ai-quest-state-driven-even-more-chinas-can-it-work/161519/, November 25, and Y. Tadjdeh, 2020, "China Threatens U.S. Primacy in Artificial Intelligence," *National Defense Magazine*, https://www.nationaldefensemagazine.org/articles/2020/10/30/china-threatens-us-primacy-in-artificial-intelligence, October 30.

ing, validating, and interpreting the sheer volume of data would be delayed and overwhelm users.

According to the National Security Commission on Artificial Intelligence, "AI is the quintessential 'dual-use' technology. The ability of a machine to perceive, evaluate, and act more quickly and accurately than a human represents a competitive advantage in any field—civilian or military."[22] As adversaries compete (and surpass) the U.S. military in this space, the adoption of AI/ML to expedite data transmission and decision-making will be increasingly vital.

Currently, decision-making involves operators manually watching data feeds, taking notes on paper, making phone calls to correlate information with other operators monitoring different data feeds, walking between computer hubs to discuss critical information, and using their own (human) analyses to provide visual and oral updates to convey this information to decision-makers. Accordingly, this approach requires massive manpower, is prone to human error, and significantly delays decisions owing to the sheer volume of complex data being communicated.[23]

To address this deficiency, the DAF Chief Architect and the Air Force Rapid Capabilities Office are using AI as an enabler and have recruited commercial companies to provide AI/ML-based analytics to transform ABMS's C2 capabilities. They have incorporated AI as part of ABMS's smartONE capability to develop algorithms for sensing and synthesizing data. In ABMS's on-ramp 2 exercise, users were able to leverage smartONE in concert with omniaONE, a common operating picture, to cue users to potentially useful information regarding an adversary's strategic assets.[24] AI was also tested and employed in on-ramp 4, where it was incorporated as part of the kill chain. Users were able to rapidly relay data between different platforms through cloudONE, the tactical-edge cloud and dataONE, ABMS's common data standardization repository. Additionally, the demonstration used AI to dial in targets to fire upon.

The eventual goal is to leverage AI/ML to provide more automation and predictive analytics to expedite data transport and decision-making even faster. The Director for Joint Force Integration in the Air Force's Strategy, Integration, and Requirements Directorate explained, "Where we are going, is to identify ways in

---

[22] NSCAI (National Security Commission on Artificial Intelligence), 2021, *National Security Commission on Artificial Intelligence Final Report*, p. 9, https://www.nscai.gov/wp-content/uploads/2021/03/Full-Report-Digital-1.pdf.

[23] See J. Eddins, 2021, "Valenzia: ABMS Will Deliver the 'Decision Advantage,'" *Airman Magazine*, https://www.macdill.af.mil/News/Features/Display/Article/2647112/valenzia-abms-will-deliver-the-decision-advantage/, May 26.

[24] M.D. Strohmeyer, 2021, "United States Northern Command Support to ABMS," Presentation to the Air Force ABMS Committee, February 24. See also V. Insinna, 2020, "Behind the Scenes of the US Air Force's Second Test of Its Game-Changing Battle Management System."

which we can take that same information and move it through the system machine to machine. So, an automated process ... to help make sense or otherwise connect the dots in a way that maybe weren't connected in the past. This is so that when that information shows up to the decision-maker, they're able to make a highly informed and fast decision."[25]

The committee views these efforts as a notable first step in incorporating AI/ML into ABMS. However, a more comprehensive expansion of AI across ABMS is needed. A commercial capability that may be considered for use is highly capable processing technologies, such as hyperautomation or intelligent process automation. Hyperautomation is "a business-driven, disciplined approach that organizations use to rapidly identify, vet and automate as many business and IT processes as possible. [It] involves the orchestrated use of multiple technologies, tools or platforms," including the following:

- AI;
- ML;
- Event-driven software architecture;
- Robotic process automation (RPA);
- Business process management (BPM) and intelligent business process management suites (iBPMS);
- Integration platform as a service (iPaaS);
- Low-code/no-code tools;
- Packaged software; and
- Other types of decision, process, and task automation tools.[26]

While intended primarily for business systems, in the context of ABMS, hyperautomation could potentially improve the accuracy of information processed and accelerate decision-making by further automating advanced C2 functions and data processing. Within the commercial sector, companies are using hyperautomation to reduce burdens on operators and increase the accuracy of predictive analytics by as much as 95 percent when trained with multiple, high-quality data sets.[27]

---

[25] J. Eddins, 2021, "Valenzia: ABMS Will Deliver the 'Decision Advantage.'"
[26] Gartner, "Hyperautomation," *Gartner Glossary*, https://www.gartner.com/en/information-technology/glossary/hyperautomation. For more on hyperautomation, see IBM Cloud Education, "What Is Hyperautomation?" *IBM Cloud Learn Hub*, https://www.ibm.com/cloud/learn/hyperautomation, April 15, 2021, and D. Wright, "Hyperautomation: The Next Digital Frontier," *Forbes*, https://www.forbes.com/sites/servicenow/2021/03/26/hyperautomation-the-next-digital-frontier/?sh=50b4919273fd, March 26.
[27] "Trends in Machine Learning to Know for 2021," *Business World Innovative Technologies*, https://www.businessworldit.com/ai/machine-learning-trends/, March 23.

To improve and maintain the quality of data collected, the process of automated machine learning (AutoML) may also be considered for adoption. AutoML is "the process of automating the time-consuming iterative task of ML model development. It allows data scientists, analysts, and developers to build ML models with high scale, efficiency, and productivity all while sustaining model quality."[28] These combined processes would enable ABMS developers and users to increase both the efficiency and efficacy of data processing; thereby, contributing to the nation's information advantage.

> **RECOMMENDATION 6: To the maximum extent possible, the Department of the Air Force Chief Architect's Office and the Department of the Air Force Rapid Capabilities Office should design and execute a comprehensive artificial intelligence strategy that would encompass all elements, to include doctrine, chain of command, policy, authorization for weapon release in a joint environment, interfaces to Joint All-Domain Command and Control, and not just select capabilities of the Advanced Battle Management System.**

Data and Data Standards

As highlighted in the *DoD's Data Strategy*, data is a strategic asset.[29] Within ABMS, data constitutes the intelligence, indications, warnings, signals, status, situation, commands, controls, and other multimodal information needed to understand the situation and command, control, and operate military forces. As such, "data in the DoD is a high-interest commodity and must be leveraged in a way that brings both immediate and lasting military advantage."[30]

The recognition of data as a strategic asset requires that data pedigree and security be maintained at all times. Original source and combined data should be tagged, catalogued, and securely stored immediately. "In a data-dependent and data-saturated world, victory belongs to the side with decision superiority—the ability to sense, make sense of a complex and adaptive environment, and act smarter, faster, and better."[31]

However, without a set of enterprise-level data standards, particularly as each military Service and DoD agency establishes its own contribution to JADC2, information sharing at scale will not be possible. The *DoD's Data Strategy* provides

---

[28] "What Is Automated Machine Learning (AutoML)?" *Microsoft Research*, https://docs.microsoft.com/en-us/azure/machine-learning/concept-automated-ml, July 1.

[29] DoD (U.S. Department of Defense), 2020, *DoD Data Strategy*, https://media.defense.gov/2020/Oct/08/2002514180/-1/-1/0/DOD-DATA-STRATEGY.PDF.

[30] DoD (U.S. Department of Defense), 2020, *DoD Data Strategy*, https://media.defense.gov/2020/Oct/08/2002514180/-1/-1/0/DOD-DATA-STRATEGY.PDF, p. 3.

[31] C. Pope, 2021, "With Its Promise and Performance Confirmed, ABMS Moves to a New Phase."

the "overarching vision, focus areas, guiding principles, essential capabilities, and goals necessary to transform the Department into a data-centric enterprise."[32] But it directs each military Service and DoD component/agency to develop its own data strategy implementation plan. This could potentially lead to misalignments and unnecessary redundancies.

The Joint Staff J6 is the DoD's lead organization for establishing common data standards for JADC2. J6 has hosted monthly meetings through its JADC2 cross functional team (CFT), composed of members from DoD agencies, military Services, the U.S. Department of Homeland Security, and NATO. The CFT held a multi-day data summit in January 2021 with the aim of developing a common data fabric[33] for JADC2 that incorporates a standard lexicon, performance metrics, and requirements for setting common data standards.[34] To date, the CFT has developed a common vocabulary and identified components of the data fabric definition that will be turned into objectives to direct subsequent work in support of JADC2. These components include metadata tagging, common data interfaces, data access control, data security, and data infrastructure.[35] Still, the common data fabric has yet to be completed owing to the challenge of establishing a standard that is neither too prescriptive, nor too open.

Additionally, in June 2021 the J6 released a classified *JADC2 Strategy* that "provides the governance and framework necessary to enable rapid integration of artificial intelligence, ML, predictive analytics and other emerging technologies."[36] It focuses on five specific lines of effort (LOEs) to include data, human enterprise, technology, nuclear command and control, and the mission partner environment.[37] More recently, the J6 is finalizing details on a classified *JADC2 Implementation*

---

[32] DoD (U.S. Department of Defense), 2020, *DoD Data Strategy*.

[33] A common data fabric refers to a set of standards and IT services that allow data to be shared among different weapon systems, different C2 networks, different organizations and services and across different levels of security. T. Hitchens, 2021, "Exclusive: 'Do-or-Die' JADC2 Summit to Crunch Common Data Standards," *Breaking Defense*, https://breakingdefense.com/2021/01/exclusive-do-or-die-jadc2-summit-to-crunch-common-data-standards/, January 12.

[34] D. Crall, 2021, "Joint All Domain Command and Control," Presentation to the Air Force ABMS Committee, March 3.

[35] A. Eversden, 2021, "Getting Away from 'Anything Goes': Military Leaders Set Data Standards for Joint War Fighting," *C4ISRNet*, https://www.c4isrnet.com/battlefield-tech/it-networks/2021/01/27/getting-away-from-anything-goes-military-leaders-set-data-standards-for-joint-war-fighting/, January 27. See also S.A. Whitehead and J.S. Wellman, 2021, "Joint All Domain Command and Control (JADC2)," Presentation to the Air Force ABMS Committee, February 5.

[36] D. Vergun, 2021, "DoD Looking for Advanced Command, Control Solution," *DoD News*, https://www.defense.gov/News/News-Stories/Article/Article/2646822/dod-looking-for-advanced-command-control-solution/, June 4.

[37] A. Eversden, 2021, "With Austin's Signature on JADC2 Strategy, Top General Says It's 'Delivery Time,'" *C4ISRNet*, https://www.c4isrnet.com/battlefield-tech/it-networks/2021/06/04/with-austins-signature-on-jadc2-strategy-top-general-says-its-delivery-time/, June 4.

Strategy that includes objectives, task transactions, milestones, service contributions (to include ABMS) and other efforts being undertaken by combatant commands and other DoD agencies. The near-term (FY 2022) focus will be on developing minimally viable products and enhancing capabilities such as DevSecOps, identity, credential, and access management (ICAM), zero trust (ZT), transport layer, and cloud.[38] To ensure common software standards, the J6 is also working with the Joint Requirements Oversight Council (JROC) to mandate open software standards for ensuring cross-Service and cross-domain compatibility and interoperability across sub-systems through the use of common software interfaces.[39] These collective efforts provide ABMS (along with Project Convergence and Project Overmatch) with a guiding foundation to support their respective development activities.

RECOMMENDATION 7: The Joint All-Domain Command and Control cross functional team should reach immediate agreement on a common data fabric and security levels of the data with data standards and tools defined at the Joint level. Without a common set of agreed upon open standards with known interface exchange requirements that do not limit innovation, the military Services risk developing incompatible and stovepiped solutions.

Containerization and Kubernetes

To enable agile development in a continuous integration/continuous delivery and deployment (CI/CD)[40] environment, containerization is the optimal solution. Containers refer to a lightweight virtual machine that can be preconfigured and uploaded to a cloud or on-premises (on-prem) environment to immediately provide both the software capability and supporting operating system libraries and dependencies necessary to entirely support the capability.[41] Expensive and time-consuming integration with different operating systems and hardware platforms are

---

[38] L.C. Williams, 2021, "Pentagon Preps JADC2 Implementation Plan," *FCW*, https://fcw.com/articles/2021/09/08/dod-jadc2-plan-implementation.aspx, September 8.

[39] T. Hitchens, 2021, "Exclusive: 'Do-or-Die' JADC2 Summit to Crunch Common Data Standards."

[40] CI/CD is a method to frequently deliver apps by introducing automation into the stages of app development. The main concepts attributed to CI/CD are continuous integration, continuous delivery, and continuous deployment. See "What is CI/CD?" *Red Hat*, https://www.redhat.com/en/topics/devops/what-is-ci-cd, 2018. See also D. Samant, 2021, "7 Rules for Faster Releases with Containerized CI/CD," *Container Journal*, https://containerjournal.com/features/7-rules-for-faster-releases-with-containerized-ci-cd/, March 29.

[41] See IBM Cloud Education, 2021, "Containerization," *IBM Cloud Learn Hub*, https://www.ibm.com/cloud/learn/containerization, June 23.

avoided by encapsulating all dependencies within the container. The production of a container with the software for CI/CD is referred to as containerization. Containerization is a well-established commercial practice and is supported by a number of government and commercial software packages that enable the construction, test and evaluation, and deployment of containers in an operational environment. This level of flexibility dramatically increases the confidence that an upgrade to specific capabilities within the context of a larger system such as ABMS can be achieved with less risk to overall mission requirements.

Within the DAF, open-source Kubernetes[42] has been widely adopted to avoid vendor lock-in and to provide resilience, security, adaptability, automation, autoscaling, and an abstraction layer. It has been successfully employed on the F-16 fighter jet and the U-2 reconnaissance aircraft to enable available on-board computing power to meet advanced system and software needs on demand.[43] According to the former DAF's Chief Software Officer (CSO), in the instance of the U-2 experimentation, "The successful combination of the U-2's legacy computer system with the modern Kubernetes software was a critical milestone for the development of software containerization on existing Air Force weapon systems."[44]

To achieve these experiments, the DAF had to create a portable technology stack that included (1) the Cloud One infrastructure layer, which provided a stable and secure common development, test, and production environment; (2) Platform One, which provided software enterprise services and hardened containers, CI/CD options, and the Istio[45] service mesh layer that provided integrated, ZT security

---

[42] Kubernetes is a portable, extensible open-source platform for managing containerized workloads and services that facilitates both declarative configuration and automation. See the Office of the Department of the U.S. Air Force Chief Software Officer, "Kubernetes," https://software.af.mil/training/kubernetes/. The key difference between containers and Kubernetes is that containers are designed to code once and run anywhere, while Kubernetes provides the potential to orchestrate and manage all container resources from a single control plane. See Microsoft, "Kubernetes vs. Docker," https://azure.microsoft.com/en-us/topic/kubernetes-vs-docker/.

[43] See S. Miller, 2020, "Why the Air Force Put Kubernetes in an F-16," https://gcn.com/articles/2020/01/07/af-kubernetes-f16.aspx, *GCN*, January 7, and K. Reichman, 2020, "In a First for the DoD, Kubernetes Installed on U-2 Dragon Lady," *Aviation Today*, https://www.aviationtoday.com/2020/10/09/first-dod-kubernetes-installed-u-2-dragon-lady/, October 9.

[44] Air Combat Command Public Affairs, 2020, "U-2 Federal Lab Achieves Flight with Kubernetes," *Air Force News*, https://www.af.mil/News/Article-Display/Article/2375297/u-2-federal-lab-achieves-flight-with-kubernetes/, October 7.

[45] Istio is a service mesh—a modernized service networking layer that provides a transparent and language-independent way to flexibly and easily automate application network functions. See Google Cloud, "What Is Istio?" https://cloud.google.com/learn/what-is-istio.

and the architecture to enable microservices;[46] and (3) an application layer, which allowed development teams to easily construct reusable modular software or microservices that leveraged hardened containers to be used across teams.[47] Figure 2.4 illustrates the layers of this technology stack.

For ABMS, more specifically, Kubernetes is used to automate security and data analysis in development. During the February 2021 multi-nation on-ramp 4 experimentation that took place in Ramstein Air Base in Germany, the combination of a Kubernetes cluster, DevSecOps (development, security, and operations), deployment of an AI/ML application at the edge, and the ability to transfer development

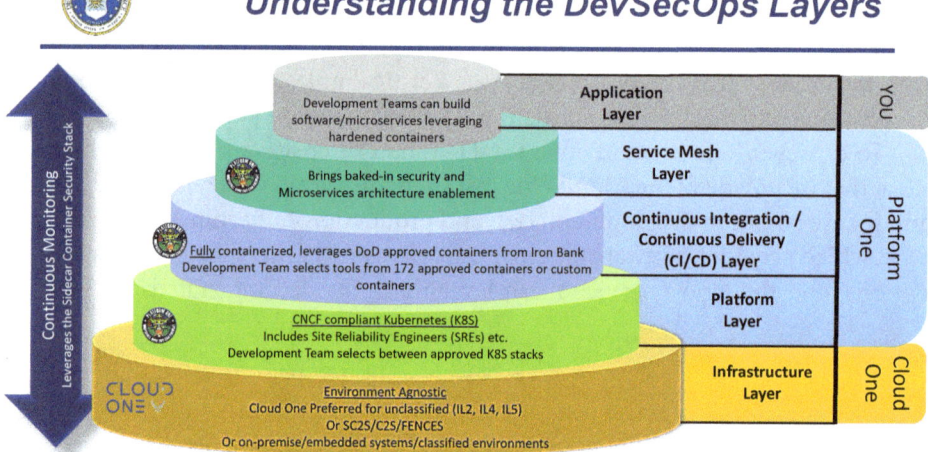

FIGURE 2.4 Understanding DevSecOps layers. SOURCE: Department of the U.S. Air Force Chief Software Officer. N. Chaillan, 2020, "DoD Enterprise DevSecOps Initiative and Platform One," Department of the U.S. Air Force Chief Software Officer, https://software.af.mil/dsop/documents/, September 15.

---

[46] Microservices (or microservices architecture) are a cloud native architectural approach in which a single application is composed of many loosely coupled and independently deployable smaller components or services. These services typically have their own technology stack, communicate with one another over a combination of REST APIs, event streaming, and message brokers; and are organized by business capability. See IBM Cloud Education, 2021, "Microservices," IBM Cloud Learn Hub, https://www.ibm.com/cloud/learn/microservices, March 30, and Amazon Web Services, "What Are Microservices?" https://aws.amazon.com/microservices/.

[47] S. Miller, 2020, "Why the Air Force Put Kubernetes in an F-16," GCN, https://gcn.com/articles/2020/01/07/af-kubernetes-f16.aspx, January 7.

code from unclassified to classified networks was successfully demonstrated.[48] The goal is to use Kubernetes to automate security by constantly scanning for anomalies and potential breaking points within ABMS. Once an anomaly or breaking point have been detected through Kubernetes, they may then be integrated within the container and deployed as any other capability advance. This obviates the need for a separate and distinct security mitigation process and fully integrates system security requirements into the DevSecOps development process. Furthermore, this approach would reduce the need for manual testing and minimize the potential for human error.

FINDING 6: Containerization and Kubernetes are mature open-source orchestration systems for enabling and securing agile development within a CI/CD environment.

RECOMMENDATION 8: In coordination with the Department of the Air Force Chief Software Officer, the Department of the Air Force Chief Architect's Office and the Department of the Air Force Rapid Capabilities Office should expand the use of containerization and Kubernetes for continuous Advanced Battle Management System development and for detecting and mitigating security vulnerabilities.

SOFTWARE CONSIDERATIONS

ABMS requires the complex integration of various independently evolving, customized weapon-specific software elements. Together, they are envisioned to support a highly distributed framework of sensing, timely aggregation, and analysis functions to provide enhanced decision-making capabilities to fight and defend with agility and resilience. Shared information with heterogeneous physical, technical, and temporal characteristics must be standardized, or at least translatable and interoperable, to facilitate timely and accurate decision-making. Coordinated and distributed actuation, defined as the application of weapons effects, are also needed to produce coherent lethal effects. Moreover, legacy systems will have to be integrated in a coherent, phased, and cost-effective manner.

The software needed to support such an integrated and multi-layered C2 framework will likely evolve as new operational challenges and adversarial strategies change over time. This will require the software to be agile, adaptable, modifiable,

---

[48] See AWS (Amazon Web Services), 2021, "Bringing Cloud Capability to the Air Force at the 'Speed of Mission Need,'" *AWS Public Sector Blog*, https://aws.amazon.com/blogs/publicsector/bringing-cloud-air-force-speed-of-mission-need/, May 7.

and secure. Rigorous testing to evaluate, verify, and improve software performance and reduce development costs will also be critical.

Application Software and DevSecOps

Modern day applications require the most up-to-date development, security, and operations (DevSecOps)[49] process, including automated security tools and automated test and deployment. Rapid advances in handheld digital devices (with integrated communication, computation capabilities, and impressive storage encrypted with biometrics and other means) have ushered in an explosive growth in innovative tools with low barriers to transform insights to prototypes and to process data at the tactical edge. Harvesting this collection of viable ideas into workable products is possible with the DevSecOps methodology, where innovators, developers, integrators, and end users all interact in frequent cycles that build upon each other.

The goal of the application environment would then be to create a model-based software engineering system that steadily combines evolving innovation, acquisition, and sustainment processes; provides telemetry and feedback from deployed code that updates the model; generates code automatically; provides automated testing and integration; and deploys the code. Any automated code development without in-depth human participation during the development process is likely to result in limited operational latitude and increased vulnerabilities. Knowledge engineering tools do exist, but must be incorporated in the software development process.

The complexity and evolving nature of ABMS requires a software development approach that is open, agile, secure, and adaptive. The DAF's CSO has taken steps to shift the department from a traditional waterfall software development approach, which typically takes anywhere from 3 to 10 years to execute, toward more agile and secure software development methodologies. In coordination with the Office of the Under Secretary of Defense for Acquisition and Sustainment (OUSD[A&S]), the DoD's Chief Information Officer (CIO), the Defense Information Systems Agency (DISA), and the other military Services, the DAF CSO has implemented the DoD Enterprise DevSecOps Initiative (DSOP) that detailed "a combination of

---

[49] DevSecOps—short for *development, security,* and *operations*—automates the integration of security at every phase of the software development life cycle, from initial design through integration, testing, deployment, and software delivery. IBM Cloud Education, 2020, "DevSecOps," *IBM Cloud Learn Hub*, https://www.ibm.com/cloud/learn/devsecops, July 30.

Kubernetes, Istio, knative,[50] and an internally developed specification for 'hardening' containers with a strict set of security requirements as the default software development platform across the military."[51] Figure 2.5 describes this initiative.

By transitioning to this enterprise approach, the DAF is able to deploy hardened software factories on existing or new environments (to include classified, disconnected, and clouds) within days instead of years; establish multiple DevSecOps pipelines with various options; enable rapid prototyping and faster deployment for weapons, C4ISR, and business systems; provide continuous learning and feedback from end users and warfighters; aid in security repairs within minutes; generate a holistic and integrated cybersecurity stack to allow complete visibility of all assets, software security state, and infrastructure as code; and facilitate adoption of microservices by using the Microservices Architecture.[52] The objective is to establish a software ecosystem with multiple innovation hubs through the CloudOne infrastructure that undergirds ABMS.[53] Figure 2.6 depicts the locations of these innovation centers that comprise the current DAF software ecosystem.

The Shadow Operations Center (ShOC-N) at Nellis Air Force Base has been designated as the lead agency responsible for creating and testing information technology applications for the ABMS.[54] The use of DevSecOps is central to ABMS's development by assembling software developers, operators, and end users in a "virtual and physical playground for information collection and sense-making using data."[55] This integration will enable quick delivery of operationally relevant and cyber secure software to troops. In June 2021, ShOC-N hosted the JADC2 21-1, J6 campaign, which convened experts from all domains and connected 17 different DoD battle laboratories to exchange operationally relevant data. Using a

---

[50] Knative is an extension of the Kubernetes container orchestration platform that enables serverless workloads to run on Kubernetes clusters, and provides tools and utilities that make building, deploying, and managing containerized applications within Kubernetes a simpler and more "native-to-Kubernetes" experience. See IBM Cloud Education, 2021, "Knative," IBM Cloud Learn Hub, https://www.ibm.com/cloud/learn/knative, June 17.

[51] T. Krazit, 2019, "How the U.S. Air Force Deployed Kubernetes and Istio on an F-16 in 45 Days," The New Stack, https://thenewstack.io/how-the-u-s-air-force-deployed-kubernetes-and-istio-on-an-f-16-in-45-days/, December 24.

[52] Office of the DAF Chief Software Officer, "DoD Enterprise DevSecOps Initiative (DSOP)," https://software.af.mil/dsop/#valuefordod.

[53] For more on Cloud One, see Office of the DAF Chief Software Officer, "Cloud One," https://software.af.mil/team/cloud-one/, and 2021, "Cloud One: Enabling Cloud for Almost Any Department of Defense Use Case," *Air Force Magazine*, https://www.airforcemag.com/cloud-one-enabling-cloud-for-almost-any-department-of-defense-use-case/, July 2.

[54] See C. Collins, 2021, "Air Force Laboratory Applies DevSecOps to Support Battle Management System Development," *ExecutiveGov Daily*, https://www.executivegov.com/2021/06/air-force-laboratory-applies-devsecops-to-support-battle-management-system-development/, June 24.

[55] "ShOC-N at Nellis Air Force Base Supports ABMS Development," *Air Force Technology*, https://www.airforce-technology.com/news/shoc-n-nellis-air-force-base-abms-development/, June 24.

FIGURE 2.5 DoD enterprise DevSecOps. SOURCE: Department of the U.S. Air Force Chief Software Officer. N. Chaillan, 2020, "DoD Enterprise DevSecOps Initiative and Platform One," Department of the U.S. Air Force Chief Software Officer, https://software.af.mil/dsop/documents/, September 15.

combination of DevSecOps and experimental software applications,[56] developers were able to assist "warfighters [to] visualize and make sense of the cyber domain and gain decision advantage over the adversary … and to enable better understanding of the cyber domain from all branches' perspectives."[57] By leveraging emerging and experimental technologies to provide real-time feedback to software engineers, this allowed them to make rapid adjustments and improvements and shorten the timeline for development.

The next step is to integrate AI/ML to enable big data processing and predictive analytics. According to the Commander, U.S. Northern Command (USNORTHCOM), the key to ensuring victory in the future all-domain battlespace is through predictive analysis. "We see JADC2 as absolutely core to the way we're

---

[56] Project IKE is a prototyping effort that is used to map networks, assess the readiness of cyber teams, and command forces in cyberspace. The project began in 2013 at DARPA under the name Plan X and was moved to the DoD's Strategic Capabilities Office (SCO) in 2019 and officially transitioned as a program under the Joint Cyber Command and Control (JCC2) program management office. See M. Pomerleau, 2021, "A Cyber Tool That Started at DARPA Moves to Cyber Command," *C4ISRNet*, https://www.c4isrnet.com/cyber/2021/04/20/a-cyber-tool-that-started-at-darpa-moves-to-cyber-command/, April 20.

[57] N. Mathison, 2021, "Nellis AFB Empowers Warfighters via DevSecOps," *Air Force News*, https://www.af.mil/News/Article-Display/Article/2668568/nellis-afb-empowers-warfighters-via-devsecops/, June 23.

Architecture and Data

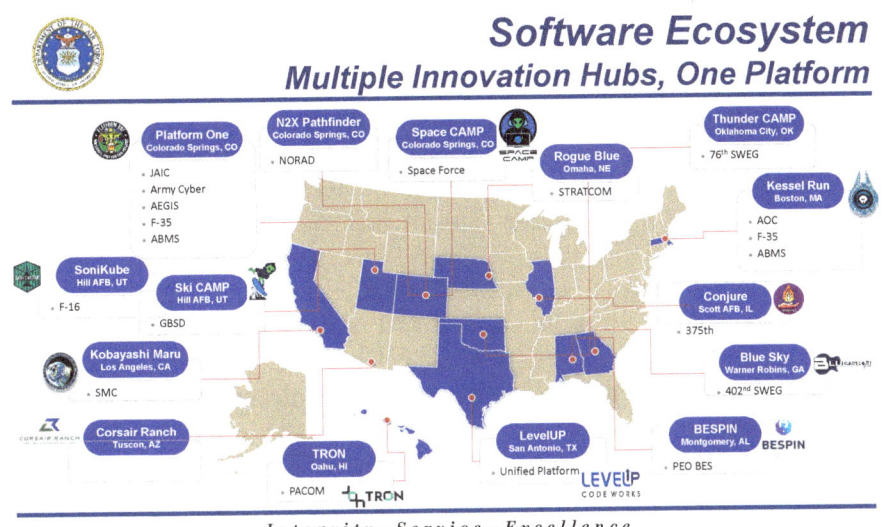

FIGURE 2.6 Software ecosystem. SOURCE: Department of the U.S. Air Force Chief Software Officer. N. Chaillan, 2020, "DoD Enterprise DevSecOps Initiative and Platform One," Department of the U.S. Air Force Chief Software Officer, https://software.af.mil/dsop/documents/, September 15.

going defend the homeland … and the part that I think is going to be so incredibly game-changing is the ability for us to really use predictive analysis and inform our decisions going into the future."[58] More specifically within ABMS, predictive analytics will enable developers to "get digital feedback on where the information is going, where the data's being corrupted, where are we having problems getting enough information through, and how is that impacting the decision making."[59]

> **FINDING 7:** The use of DevSecOps within the existing software factories has been effective at creating a CI/CD environment for some ABMS development.

> **RECOMMENDATION 9:** The Department of the Air Force Chief Architect's Office and the Department of the Air Force Rapid Capabilities Office should adopt development, security, and operations as the common development environment using containerization and continuous integration/continuous delivery across all of the Advanced Battle Management System.

---

[58] T. Hitchens, 2020, "The Key to All-Domain Warfare Is 'Predictive Analysis;' Gen. O'Shaughnessy," Breaking Defense, https://breakingdefense.com/2020/05/the-key-to-all-domain-warfare-is-predictive-analysis-gen-oshaughnessy/, May 5.

[59] N. Mathison, 2020, "Nellis AFB Empowers Warfighters via DevSecOps."

Data Rights

In defense acquisition, the tendency is for the government to ask industry for access to all data rights to include both technical and computer software data. Technical data includes "any recorded information of a scientific or technical nature (e.g., product design or maintenance data, computer databases, and computer software documentation").[60] Computer software includes the "executable code, source code, code listing, design details, processes, flow charts, and related material that would enable the software to be reproduced, recreated, or recompiled."[61] Industry commonly retains title to technical data and computer software, but conveys a licensing arrangement for government agencies to use such data through one of three categories: unlimited rights, limited rights (technical data) or restricted rights (computer software), and government purpose rights.[62]

For ABMS, where the architecture is comprised of an amalgam of hardware, software, legacy systems, and other C2 infrastructure, owning or accessing data rights may not be feasible nor sustainable. Instead, focus should be given to owning the rights (instead of all intellectual property rights) to the interfaces that connect these various components.[63] This is particularly relevant for systems using open system architectures (OSA). "A major benefit of the modular/open systems approach from a data rights perspective is that the performance and interface information should meet the DFARS [Defense Federal Acquisition Regulation Supplement] criteria of form, fit, and function data and thereby have no government data rights restrictions."[64]

By acquiring and maintaining the performance and interface requirements, the government would be able to use them to procure and support modular solutions

---

[60] DISA (Defense Information Systems Agency), "Data Rights," https://disa.mil/about/legal-and-regulatory/datarights-ip/datarights, accessed September 18, 2021.

[61] DISA (Defense Information Systems Agency), "Data Rights," https://disa.mil/about/legal-and-regulatory/datarights-ip/datarights, accessed September 18, 2021. See also Carnegie Mellon University, "Governing Rights in Technical Data and Computer Software," https://www.cmu.edu/osp/contracts/contracts-process/rights.html, accessed September 18, 2021.

[62] See W.J. DeVecchio, 2018, "Taking the Mystery Out of Data Rights," *Thomson Reuters*, Issue 18-8, https://media2.mofo.com/documents/180700-mystery-data-rights.pdf, July, and S.B. Cassidy, A.B. Hastings, and J.L. Plitsch, 2017, "What Every Company Should Know About IP Rights When Selling to the US Government," *Landslide*, 9(6), July–August.

[63] See U.S. Department of Defense Instruction 5010.44, 2019, "Intellectual Property (IP) Acquisition and Licensing," Office of the Under Secretary of Defense for Acquisition and Sustainment, https://www.esd.whs.mil/Portals/54/Documents/DD/issuances/dodi/501044p.PDF, accessed November 26, 2021.

[64] Office of the Assistant Secretary of the Army for Acquisition, Technology, and Logistics, 2015, "Army Data and Data Rights (D&DR) Guide," p. 35, https://www.acq.osd.mil/dpap/cpic/cp/docs/Army_Data_and_Data_Rights_Guide_1st_Edition_4_Aug_2015.pdf, August.

from multiple vendors. This would in turn reduce the dependency on sole source acquisitions or costly procurements of licensing rights to privately developed technology or computer software. "While any weapon system can potentially benefit from an OSA approach, sub-systems expected to contain proprietary technology, have frequent technology updates, or are available from multiple sources, are particularly strong candidates. Another reason to consider the modular/open system approach would be if the sub-system were expected to have data rights restrictions, which are not likely to be mitigated through the acquisition of additional data rights."[65]

RECOMMENDATION 10: For modular open system designs with robust interface specifications, the Department of the Air Force Rapid Capabilities Office should acquire performance and interface requirements instead of all intellectual property rights.

SECURITY

Network Reliability, Resiliency, and Fault Tolerance

The success of ABMS during any engagement relies on the communications infrastructure. Much of this infrastructure consists of data networks that are responsible for transporting essential data from sensors to computation, decision-making, execution, and data storage nodes. As a critical element of ABMS, it is important to monitor real-time status of the state and resilience of the entire network infrastructure at all times. Should there be any disruption or degradation of network performance, redundant and resilient systems must automatically reroute and reconstitute network communications to maintain ABMS capabilities. The ability to operate with little or no communication connectivity or bandwidth is also important and should be considered as part of ABMS's overall security design.

The defined performance of the network is often structured around real-time measures of bandwidth and latency—both within individual point-to-point links of the network and across end-to-end communications. The specific acceptable metrics for bandwidth and latency are defined in the context of the tactical status of ABMS capabilities. During lower-level engagements, some degradation of bandwidth and latency may be acceptable. However, during higher-level engagements, where cyberattacks are more likely, bandwidth must be elevated and latency reduced, or at a minimal, maintained. Thus, the question of when the network can no longer support volume and timeliness of priority traffic is a dynamic question.

---

[65] Ibid.

It must also reflect the degree to which it serves the overall ABMS goal of enabling the DAF to operate within the adversary's observe-orient-decide-act (OODA) loop.

The dynamic nature of the network's minimum performance requirements creates a new set of challenges for when redundant and resilient communications structures are employed. Although the minimum performance is dynamic, it is not ad hoc, but predefined based on the C2 requirements of ABMS during engagement. Once defined, ABMS capabilities will be able to automatically adjust network capabilities (e.g., using software defined networks and network functions virtualization) to maintain a minimum performance profile at all times.

Defining minimum performance characteristics for each operational scenario will define how the network breaks and what the contingency will be for each scenario. It is important to note that the source of network failure may be forced by the adversary and resilient and redundant capabilities may be anticipated by the adversary, as well. Any contingency must therefore realistically address the threat model, where the adversary will directly attack the network and communications infrastructure necessary to enable ABMS.

As a contributor to both JADC2 and the nation's NC3 capability, security, reliability, and resiliency should be essential attributes of ABMS. According to Representative Adam Smith (D-WA), Chairman of the House Armed Services Committee, "We have to be able to protect [our command and control] systems and ideally we have to be able to build a system so that we can make our adversary systems more vulnerable. That really needs to be the focus."[66] The Vice Chairman of the Joint Chiefs of Staff added, "It's important to realize that JADC2 and NC3 are intertwined because … NC3 will operate in significant elements of JADC2. Therefore, NC3 has to inform JADC2 and JADC2 has to inform NC3. You have to have that interface back and forth, and that's been recognized."[67] As a C2 system that is able to bridge different security classifications through crossdomainONE,[68] ABMS will need to provide the secure, reliable, and resilient interfaces to connect JADC2 and NC3. This will require the participation of the U.S. Strategic Command (USSTRATCOM).

As detailed in Chapter 1, current Air Operations Center (AOC) modules were designed and based on now-dated technologies—including those used for cybersecurity. The incremental development and evolution of safeguarding AOC compo-

---

[66] L.C. Williams, 2021, "JADC2 Needs to Be Pentagon's 'Big Bet,' Flournoy Says," *FCW*, https://fcw.com/articles/2021/03/30/jadc2-big-bet-flournoy.aspx, March 30.

[67] C. Clark, 2020, "Nuclear C3 Goes All Domain: Gen. Hyten," *Breaking Defense*, https://breakingdefense.com/2020/02/nuclear-c3-goes-all-domain-gen-hyten/, February 20.

[68] CrossdomainONE is a platform that would seamlessly and securely move data up and down security classification boundaries. See T. Hitchens, 2019, "First Multi Domain C2 Exercise Planned: 'ABMS Onramp,' " *Breaking Defense*, https://breakingdefense.com/2019/12/first-multi-domain-c2-exercise-planned-cross-domain-one/, December 6.

nents resulted in a patchwork of varying constructs, processes, and completeness. Furthermore, the design and development of the current AOC were executed in the context of a former threat environment—one far different from what operators encounter today. The battlespaces of today and the future are highly complex and will engage all domains at the strategic, operational, and tactical levels. What is most difficult to anticipate and prepare for is competition conducted in the "gray zone" between peace and war.[69] The unpredictability and complexity of all-domain warfare require confidentiality, integrity, and availability of the operational environment in which ABMS supports.

To accomplish this, a security engineering construct and associated implementation strategy from a multi-pronged approach to consider the operational, technical, and risk assessment (based on a realistic assessment of the threat environment and technical limitations) perspectives should be developed. The security construct should identify vulnerabilities and support:

- *Security:* ABMS must provide secure and accurate data and be trusted by operators across all domains;
- *Reliability:* ABMS capabilities must perform as advertised and made available, when needed, and;
- *Resilience:* ABMS must be able to perform in anti-access and degraded environments and be able to rapidly reconstitute, as needed.

Ultimately, security engineering must be designed, implemented, and evaluated against the current operational and technical baseline. It cannot be solely a technical evaluation.

FINDING 8: ABMS—and the operational forces that use it—must be resilient to technical failures and limitations plus against adversarial attacks.

FINDING 9: ABMS bandwidth and latency capacities need to adjust to changing operating conditions and demands. For example, during high-level engagements, where cyberattacks are more likely, bandwidth must be elevated and latency reduced or at least maintained or in support of critical functions,

---

[69] "Gray zone" refers to competitive interactions among and within state and non-state actors that fall between traditional war and peace. It is characterized by ambiguity about the nature of the conflict, opacity of the parties involved, or uncertainty about the relevant policy and legal frameworks. Adversaries seek competitive advantages through military, diplomatic, information, and economic tactics. See J.L. Votel, C.T. Cleveland, C.T. Connett, and W. Irwin, 2016, "Unconventional Warfare in the Gray Zone," *Joint Forces Quarterly*, 80(1st quarter) 101–109. See also L.J. Morris, M.J. Mazarr, J.W. Hornung, S. Pezard, A. Binnendijk, and M. Kepe, 2019, "Gaining Competitive Advantage in the Gray Zone," RAND Corporation, https://www.rand.org/pubs/research_reports/RR2942.html.

associated data transfer need to be prioritized, where bandwidth availability is challenged.

RECOMMENDATION 11: The Department of the Air Force Chief Architect's Office and the Department of the Air Force Rapid Capabilities Office should design resilience into the Advanced Battle Management System architecture and specify dynamic criteria for needed performance.

Multi-Level Security

Multi-level security (MLS) refers to "processing information with different classifications and categories that simultaneously permits access by users with different security clearances and denies access to users who lack authorization."[70] Such systems would enable individual platforms to process information at different security levels and move data between these levels, as appropriate.[71] The advantages of MLS are to store and share data of mixed classification and to provide secure access to multiple classifications of data to those who have the proper authorizations.

ABMS has incorporated this MLS approach through CrossDomainONE, the platform that enables data to move across classification levels and the ABMS DeviceOne SecureView (ADSV) established by the Air Force Research Laboratory (AFRL).[72] ADSV supports ABMS and JADC2 by connecting sensors and shooters and enabling data transfer and/or access to both multi- and cross-domain data. Figure 2.7 summarizes the key features of ADSV.

The committee views CrossDomainONE and ADSV as notable first steps in enhancing MLS protection for ABMS. However, ADSV's use of hypervisor—a virtual machine monitor—for security enhancement may expose ABMS to potential vulnerabilities and risks. A critical next step is to automate data transfer and integration. Data from sensors and platforms, particularly for highly classified intelligence and sensor data, are rarely meta-tagged (labeled) to allow a MLS system

---

[70] Information Technology Laboratory Computer Security Resource Center, "Multi-Level Security (MLS)," National Institute of Standards and Technology (NIST), https://csrc.nist.gov/glossary/term/multi_level_security.

[71] For more on MLS, see E. Boebert, 2008, "Multilevel Security," pp. 239–273 in *Security Engineering: A Guide to Building Dependable Distributed Systems, Second Edition* (R. Anderson, ed.), Wiley, http://www.cse.psu.edu/~pdm12/cse597g-f15/readings/cse597g-mls_reading.pdf.

[72] SecureView® cross domain access solution allows users access to multiple independent levels of security (MILS) on a single workstation and provides immediate access to mission critical data. It was developed by the Air Force Research Laboratory to provide the Intelligence Community (IC) with unparalleled security and protection against data exfiltration. See AFRL (Air Force Research Laboratory), "Advanced Battle Management System (ABMS) DeviceOne SecureView," https://afresearchlab.com/technology/successstories/advanced-battle-management-system-abms-deviceone-secureview/.

Capability

- Supports NIPRNet, SIPRNet, and Coalition access on a single PC or laptop
- Intuitive user interface that requires minimal training for end users
- Enables secure mobility solutions for Executive Communication and traveling personnel
- Seamlessly supports high performance and high-bandwidth applications

Security

- Type I bare-metal hypervisor enhances the cyber defense posture of government workstations
- Minimizes Type I encryptors by integrating support for NSA's Commercial Solutions for Classified (CSfC)
- Ensures 100% Trusted Boot and Secure Isolation in the hardware
- Has received highest MILS evaluation to date against NIST 800-53 Criteria

Flexibility

- Supports either Standard Desktop Configuration (SDC) or Thin Virtual Desktop Infrastructure (VDI)
- Enables rapid provisioning, management, and re-configuration of workstations
- Government off-the-shelf (GOTS) solution based on OpenXT which meets DoD's open-source requirement

FIGURE 2.7 ABMS DeviceOne SecureView®. SOURCE: Air Force Research Laboratory Information Directorate, SecureView Cross Domain Access Solution.

to effectively evaluate a machine-to-machine transport and integration of data. To address this technical gap, platforms and systems should be able to operate at different classification levels to ensure both compatibility with other Services and multi-national partners and allies operating in the JADC2 framework. Sanitizing subsets of highly classified information for release to operational users should also be addressed in MLS design.

Cybersecurity and Zero Trust

Security is critical to ABMS. As such, the storage, processing, and communication of data must be equally secure and reliable, and communication must be enabled and timely across the enterprise. Designing a holistic cybersecurity architecture for all of ABMS and its supporting components is essential; piecemeal development will only result in security gaps and seams that could (and will) be exploited by our adversaries. According to the Deputy Commander of the 16th Air Force/Air Forces Cyber (the Air Force's information warfare command), "The only way we're going to be able to really conduct JADC2 is through a defended, resilient, fully capable fabric, warfighting communication fabric ... we're going to have to not just enable that and design it and operate it, we're going to have to defend it because the adversary is going to try to take that away from us."[73]

To address this security challenge, ABMS developers are moving toward the adoption of zero trust (ZT), an architecture that provides authenticated and authorized access between services without relying on the location of those services within a network infrastructure.[74] ZT is not a specific set of technologies, but rather an architectural construct that requires authentication and authorization for all interactions between individual nodes in a distributed system. Most commonly, ZT relates to deperimeterization (also, de-perimeterization), the ability to protect an organization's systems and data on multiple levels through a mix of encryption, secure computer protocols, secure computer systems, and data-level authentication. Deperimeterization reduces or removes the need to operate within an enterprise

---

[73] M. Pomerleau, 2020, "Air Force Looking at How to Defend JADC2 Systems," *DefenseNews*, https://www.defensenews.com/digital-show-dailies/air-force-association/2020/09/16/air-force-looking-at-how-to-defend-jadc2-systems/, September 16.

[74] See S. Rose, O. Borchert, S. Mitchell, and S. Connelly, 2020, "Zero Trust Architecture," NIST Special Publication 800-207, https://nvlpubs.nist.gov/nistpubs/SpecialPublications/NIST.SP.800-207.pdf, August. For more on zero trust in the federal government, see S. Vetter and A. Stewart, 2020, "Zero Trust: Evolving the Federal Government's Security Model!" *PSC Magazine*, https://www.pscouncil.org/a/Content/2020/Zero_Trust__Evolving_the_Federal_Government_s_Security_Model.aspx. See also, C. Clark, 2020, "Dunlap on Zero Trust, Agility, and ADO Cybersecurity," *Breaking Defense*, https://breakingdefense.com/2020/11/dunlap-on-zero-trust-agility-ado-cybersecurity/, November 24.

network by providing access between a user and distributed services such as data, specialized computing, sensors, and location services.[75]

For ABMS, ZT is being used on deviceONE, which certifies credentials are checked at every layer of the system. It is important for the DAF Chief Architect and the Air Force Rapid Capabilities Office to expand the use of ZT and other cybersecurity protections across all ABMS capabilities to ensure maximum safeguards against cyber vulnerabilities.

While the concept of ZT has been in existence for a while, recent improvements in support technologies to connect a wide variety of services through multi-factor authentication (MFA), authorization services, and other security capabilities have enabled wider adoption of ZT. ZT provides authenticated and authorized access between dispersed platforms and users to enable command, control, and communications within a dynamic network environment. It is also frequently used in cloud computing environments, which is suitable for ABMS's Internet of Military Things, to provide trusted access from endpoints in untrustworthy network environments to a cloud computing and data environment.

However, adopting ZT architecture not only for ABMS, but across the entire DoD JADC2 enterprise may be fraught with challenges. ZT technology that is able to address the enormity of DoD-wide C4ISR in support of the spectrum of military operations has not been proven sufficiently robust. Moreover, the wide variety of capabilities and requirements that necessitate coordination across the JWC demands creating an authorization service that would remain current while connecting all users, data, and technical platforms. The forthcoming *JADC2 Implementation Strategy* does encourage focusing greater attention on ZT, but it falls short in addressing the technical and logistics challenges of adopting ZT across disjointed and disconnected users and systems that comprise the DoD enterprise. The strategy also does not address specific authorities needed to access and actuate systems in JADC, which is essential in human-machine communications. Thus, while ZT should remain an integral component of ABMS, the overall ABMS (and JADC2) security architecture will require well-defined authentication and authorization requirements between platforms and the human-computer interface (HCI).

Another consideration is establishing cybersecurity for legacy systems that possess varying and irregular degrees of vulnerabilities owing to their outdated hardware, software, and operating technologies. These systems are generally incompatible with security features surrounding access, including MFA, single-sign

---

[75] For more on deperimeterization, see SC Staff, 2004, "Jericho Forum Brings Its Deperimeter Concept to U.S.," SC Media, https://www.scmagazine.com/news/-/jericho-forum-brings-its-deperimeterization-concept-to-u-s, July 30, and J. Kindervag, 2016, "No More Chewy Centers: The Zero Trust Model of Information Security," *Forrester*, https://crystaltechnologies.com/wp-content/uploads/2017/12/forrester-zero-trust-model-information-security.pdf, March 23.

on, and role-based access. They also lack sufficient encryption techniques necessary for protecting digital data confidentiality. Establishing connections with these legacy systems through the rubric of ABMS will require at a minimum, performing vulnerability scans to identify and address all vulnerability gaps, adopting cybersecurity controls for systems that are networked with ABMS components, and updating the latest system patches to minimize exposure.[76]

It is important to recognize that while ZT provides one layer of security protection, a comprehensive cybersecurity plan for the entire ABMS architecture is warranted.[77] This requires both offensive and defense planning to include red teaming ABMS's cyber defenses to ensure their resiliency to protect against adversarial penetrations and attacks. The Air Force's Mission Defense Teams, specialized defensive cyber teams tasked to protect critical Air Force missions and installations, should be leveraged.[78] Additionally, Congress has recently directed and authorized defense cyber personnel to operate outside of U.S. networks. This measure will enable U.S. Cyber Command (USCYBERCOM) to better understand the types of malware adversaries are employing and the types of operations they might be planning against U.S. networks through its Hunt Forward missions.[79] According to the Commander of USCYBERCOM, "We cannot afford to wait for cyberattacks

---

[76] See S. Crozier Cox and H. Levinson, 2019, "Cybersecurity Engineering for Legacy Systems: 6 Recommendations," Carnegie Mellon University Software Engineering Institute, https://insights.sei.cmu.edu/blog/cybersecurity-engineering-for-legacy-systems-6-recommendations/, August 26, and D. Snyder, J.D. Powers, E. Bodine-Baron, B. Fox, L. Kendrick, and M.H. Powell, 2015, "Improving the Cybersecurity of U.S. Air Force Military Systems Throughout Their Life Cycles," RAND Corporation, https://www.rand.org/content/dam/rand/pubs/research_reports/RR1000/RR1007/RAND_RR1007.pdf.

[77] See D. Snyder, J.D. Powers, E. Bodine-Baron, B. Fox, L. Kendrick, and M.H. Powell, 2015, "Improving the Cybersecurity of U.S. Air Force Military Systems Throughout Their Life Cycles," and 2021, "Top 11 Most Powerful Cybersecurity Software Tools in 2021," *Software Testing Help*, https://www.softwaretestinghelp.com/cybersecurity-software-tools/, October 4.

[78] See H. Stevens, 2019, "Mission Defense Team: Defending the RPA network," *Air Combat Command News*, https://www.acc.af.mil/News/Article-Display/Article/1986201/mission-defense-team-defending-the-rpa-network/, October 10, and M. Pomerleau, 2019, "When Malware Hits an F-16, Call These New Air Force Cyber Teams," *C4ISRNet*, https://www.c4isrnet.com/dod/air-force/2019/04/17/when-malware-hits-an-f-16-call-these-new-air-force-cyber-teams/, April 17.

[79] Hunt Forward missions are U.S. cyber protection teams, who operate at the request of host nations to partner with them in conducting defensive cyber operations on their (host nation) networks. See E. Tucker, 2020, "Military's Top Cyber Official Defends More Aggressive Stance," *Military Times*, https://www.militarytimes.com/news/your-military/2020/08/25/militarys-top-cyber-official-defends-more-aggressive-stance/, August 25. See also U.S. Cyber Command, 2020, "United States Cyber Command Technical Challenge Problem Set," https://www.cybercom.mil/Portals/56/Documents/2020%20Tech%20Challenge%20Problems%20UNCLASS%20CAO-PAO%20FINAL.pdf?ver=2020-08-18-160721-850.

to affect our military networks. We learned that defending our military networks requires executing operations outside our military networks. The threat evolved, and we evolved to meet it."[80]

FINDING 10: Offensive and defensive cybersecurity protection is essential for ABMS and must be holistically and seamlessly integrated into the entire system architecture from the start. Approaching cybersecurity in a fragmented approach or as an afterthought will only generate vulnerability gaps that will be exploited by malicious actors.

FINDING 11: ZT holds promise, but is not currently sufficiently mature to be the singular security protection in military C2 systems like ABMS.

RECOMMENDATION 12: The Joint Staff's J6, the Department of the Air Force, and the broader U.S. Department of Defense community should establish and implement a robust enterprise-wide offensive and defensive cybersecurity strategy for Joint All-Domain Command and Control (JADC2) and the Advanced Battle Management System. Security is a fundamental requirement that must be designed and fully integrated into the all JADC2-supporting systems' architecture from the start.

RECOMMENDATION 13: The Department of the Air Force Rapid Capabilities Office should apply zero trust (ZT) in stages as technologies mature and integrate ZT services to include the use of multi-factor authentication across all of the Advanced Battle Management System.

RECOMMENDATION 14: In addition to adopting zero trust, the Department of the Air Force Rapid Capabilities Office should leverage the best available mature cybersecurity practices and capabilities, including multi-factor authentication; identity, credential, and access management; encryption; penetration testing; managed detection services; behavior monitoring applications; among others.

RECOMMENDATION 15: The Department of the Air Force Rapid Capabilities Office (DAF RCO) should employ the Air Force's Mission Defense Teams to red team the Advanced Battle Management System's cyber defenses against attacks from malicious actors. Based on these red team

---

[80] P.M. Nakasone and M. Sulmeyer, 2020, "How to Compete in Cyberspace," *Foreign Affairs*, https://www.foreignaffairs.com/articles/united-states/2020-08-25/cybersecurity, August 25.

exercises, the DAF RCO should address vulnerabilities by bolstering and enhancing cyber defenses accordingly.

RECOMMENDATION 16: The Department of the Air Force Chief Architect's Office and the Department of the Air Force Rapid Capabilities Office should work in partnership with the U.S. Cyber Command to address Internet of Things defense and other cyber vulnerabilities and exploits that are highlighted in the "United States Cyber Command Technical Challenge Problem Set" document.

TESTING AND MODELING

Test and Evaluation

Risk reduction is the primary objective for testing and evaluation (T&E). By assessing system performance against warfighter and operator requirements, early detection and mitigation of system vulnerabilities, deficiencies, and performance issues may be accomplished. As part of the broader systems engineering process (SEP), the committee deems T&E as essential and should be conducted throughout ABMS development and deployment cycles, particularly as new products and services are introduced into the overall ecosystem.

According to the Chief Architect of the DAF, ABMS is designed to be evolving to incorporate and utilize new technologies as they emerge. Developmental testing is thus integrated into the broader DevSecOps process, while operational testing is conducted through large scale on-ramp demonstrations and exercises.[81] The first field test of ABMS took place from December 16–18, 2019, and involved the Air Force, Navy, and Army operating under a homeland defense scenario. The on-ramp exercise tested new software, communications equipment, and a mesh network to transmit time-sensitive information between fighters, destroyers, ground forces, and command centers. "Today's demo is our first time demonstrating internet-of-things connectivity across the joint force. Cloud, mesh networking and software-defined systems were the stars of the show, all developed at commercial Internet speeds ... the goal is to move quickly and deliver quickly."[82]

Subsequent operational field tests occurred from August 31 to September 3, 2020 (on-ramp 2 that focused on classified and unclassified communications and testing of 28 ABMS "ONE" product lines to connect sensors to weapons through

---

[81] P. Dunlap, 2020, "ABMS Overview," Presentation to the Air Force ABMS Committee, October 30.

[82] See C. Bousie and C. Pope, 2019, "Military Conducts First Test of Advanced Battle Management System," *Air Force News*, https://www.eglin.af.mil/News/Article-Display/Article/2047058/military-conducts-first-test-of-advanced-battle-management-system/, December 26.

a secure data network);[83] September 15 to September 25, 2020 (on-ramp 3 that tested integrating non-traditional battle management C2 nodes, such as the KC-46 tanker, with ABMS to provide seamless detection, tracking, and engagement in all domains);[84] and late February 2021 (on-ramp 4 that tested and observed the ability of the joint force, allies and multi-national partners to integrate and provide command and control across multiple networks to multiple force capabilities).[85] An on-ramp 5 was originally planned for the Pacific Theater, but was canceled in March 2021 owing to budget constraints.[86]

Additionally, Air Force and Army leaders signed a 2-year inter-Service agreement in September 2020 to work jointly to establish mutual standards for data sharing and service interfacing for ensuring all new communication equipment, networks, and AI are compatible with one another. The agreement also impacts joint force training, exercises, and demonstrations.[87]

While these activities are noteworthy, recent reductions in the ABMS budget are hampering the DAF's ability to continue with large scale live exercises and capability demonstrations. Additionally, the Secretary of the Air Force has directed focusing more attention on making "true" operational improvements vice prototyping and experimentation.[88] For these reasons, the committee determines that ABMS may need to retailor future testing through a combination of model-based systems engineering (MBSE), modeling and simulation (M&S), and the use of digital twins.

FINDING 12: Developmental testing should be focused on detecting and remedying errors in designing or implementing a system; operational testing

---

[83] D. Henley, 2020, "Advanced Battle Management System OnRamp #2, Accelerating Data-Sharing and Decision-Making," *Air Force News*, https://www.acc.af.mil/News/Article-Display/Article/2358597/advanced-battle-management-system-onramp-2-accelerating-data-sharing-and-decisi/, September 22.

[84] D. Henley, 2020, "KC-46 Tests Command and Control During ABMS Onramp 3," *Air Force News*, https://www.acc.af.mil/News/Article-Display/Article/2413971/kc-46-tests-command-and-control-during-abms-onramp-3/, November 12.

[85] B.W. Everstine, 2021, "USAFE's ABMS On-Ramp Included Partner Nations, Base Defense Scenario," *Air Force Magazine*, https://www.airforcemag.com/usafes-abms-on-ramp-included-partner-nations-base-defense-scenario/, March 1, and U.S. Air Forces in Europe and Air Force Africa, 2021, "USAFE Completes CJADC2 Demonstration," *USAFE Press Release*, https://www.usafe.af.mil/News/Press-Releases/Article/2518755/usafe-completes-cjadc2-demonstration/, March 1.

[86] B.W. Everstine, 2021, "Pacific ABMS On-Ramp Canceled Due to Budget Cuts," *Air Force Magazine*, https://www.airforcemag.com/pacific-abms-on-ramp-cancelled-due-to-budget-cuts/, March 17.

[87] G. Reim, 2020, "US Army and USAF to Jointly Develop Battlefield Network, Called, called CJADC2," *Flight Global*, https://www.flightglobal.com/defence/us-army-and-usaf-to-jointly-develop-battlefield-network-called-cjadc2/140450.article, October 2.

[88] See T. Hitchens, 2021, "Air Force ABMS Refocus: Capabilities and Kit, Not Experiments," *Breaking Defense*, https://breakingdefense.com/2021/09/air-force-abms-refocus-capabilities-and-kit-not-experiments/, September 20.

should be focused on the ability of the system to meet the user's requirements. Both developmental and operational testing—augmented with digital and MBSE, M&S, and digital twins—must be continuously executed and evaluated across ABMS to maintain the DAF's technological advantage in an increasingly sophisticated threat environment.

Model-Based Systems Engineering

In 2016, the DoD introduced "Systems Engineering Digital Engineering Fundamentals" to encourage the use of MBSE practices within a digital ecosystem.[89] MBSE is defined as the "formalized application of modeling to support system requirements, design, analysis, verification and validation activities beginning in the conceptual design phase and continuing throughout development and later life cycle phases."[90] More simply, MBSE is used to support requirements, design, analysis, verification, and validation associated with the development of complex systems.[91] The advantages of using this methodology are to reduce development risk, improve system performance, institutionalize rigor and precision into the design process, and enhance knowledge transfer. Unlike traditional systems engineering methods that focus primarily on design documentation (Document-Intensive Systems Engineering), MBSE instead focuses complex systems like ABMS that are suitable in digital-modeling environments.

As a multi-disciplinary approach, MBSE combines modeling, systems thinking, and systems engineering into a holistic framework. It covers four essential systems engineering domains to include requirements/capabilities; behavior; architecture/structure; and verification and validation. If applied to ABMS, MBSE would enable DAF engineering teams to better understand design change impacts; communicate design intent; verify, change, accept, and sustain functional capabilities; and analyze a system's design before it is constructed. Moreover, as the methodology continues to mature, MBSE may extend beyond basic engineering models to support cross-domain model integration and complex predictive and effects-based modeling.[92]

---

[89] See DoD Digital Engineering Working Group, 2016, "Systems Engineering Digital Engineering Fundamentals," https://ac.cto.mil/wp-content/uploads/2019/06/DE-Fundamentals.pdf.

[90] INCOSE (International Council on Systems Engineering Technical Operations), 2007, "Systems Engineering Vision 2020," International Council on Systems Engineering, p. 15, September.

[91] See N. Shevchenko, 2020, "An Introduction to Model-Based Systems Engineering (MBSE)," Carnegie Mellon University Software Engineering Institute, https://insights.sei.cmu.edu/blog/introduction-model-based-systems-engineering-mbse/, December 21, and K. Henderson and A. Salado, 2020, "Value and Benefits of Model-Based Systems Engineering (MBSE): Evidence from the Literature," *Wiley Periodicals LLC*, 2021(24):51–66, https://onlinelibrary.wiley.com/doi/epdf/10.1002/sys.21566, December 15.

[92] INCOSE, "Systems Engineering Vision 2020," p. 24.

Within the DAF, there has been a strong push to move toward enterprise adoption of digital engineering (DE), of which MBSE is a subset.[93] The former Assistant Secretary of the Air Force for Acquisition, Technology, and Logistics (SAF/AQ) encouraged the application of digital engineering to facilitate ownership of the technical baseline through "owning the technical stack."[94] The objectives are to:

- Achieve dominant capabilities while controlling life cycle costs;
- Increase the use of prototyping and experimentation;
- Improve requirements definition;
- Strengthen the DAF's organic engineering capability; and
- Improve DAF leaders' ability to understand and mitigate technical risk.[95]

The former SAF/AQ states that digital engineering has enabled the DAF to embrace computer-centric design and testing to expedite prototyping and reduce costs on new systems like next generation air dominance (NGAD) fighters.[96] DE is also being employed in the ground-based strategic deterrent (GBSD) system, which is pursuing a flexible design based on intelligence assessments and technology forecasts, and in Space and Missile Systems Command's protected anti-jam tactical satellite communications (PATS) family of systems that is leveraging both static and dynamic models in testing, integration, and design.[97]

Additionally in June 2021, the DAF stood up a Digital Transformation Office (DTO) headquartered at Air Force Materiel Command (AFMC) that is responsible for developing a digital governance structure and managing current and new digital

---

[93] See R. Tsui, D. Davis, and J. Sahlin, 2018, "Digital Engineering Models of Complex Systems Using Model-Based Systems Engineering (MBSE) from Enterprise Architecture (EA) to Systems of Systems (SoS) Architectures and Systems Development Life Cycle (SDLC)," 28th INCOSE International Symposium, July 7–12.

[94] W. Roper, 2020, "There Is No Spoon: The New Digital Acquisition Strategy," p. 5, https://software.af.mil/wp-content/uploads/2020/10/There-Is-No-Spoon-Digital-Acquisition-7-Oct-2020-digital-version.pdf, October 7.

[95] AFMC (Air Force Materiel Command), "Benefits of Digital Engineering," Air Force Digital Campaign, https://wss.apan.org/af/aflcmc/Benefits%20of%20DE%20Page/Home.aspx, accessed September 15.

[96] See R.S. Cohen, 2021, "ABMS, Digital Engineering Decisions on Roper's Final To-Do List," *Air Force Magazine*, https://www.airforcemag.com/abms-digital-engineering-decisions-on-ropers-final-to-do-list/, January 14.

[97] AFMC, 2020, "AF Digital Campaign Industry Exchange Day," Air Force Digital Campaign, https://youtu.be/26lot6gv3Xk, accessed September 15, 2021. For more on GBSD, see CRS, 2020, "Defense Primer: Ground Based Strategic Deterrent (GBSD) Capabilities," Congressional Research Service In Focus, https://sgp.fas.org/crs/natsec/IF11681.pdf, November 10. For more on PATS, see F. Wolfe, 2021, "Protected Anti-jam Tactical SATCOM Marquee Effort for SMC," https://www.defensedaily.com/protected-anti-jam-tactical-satcom-marquee-effort-smc/space/, February 26.

transformation activities. The DTO director is also dual-hatted as the director of the DAF's Digital Engineering Enterprise Office, responsible for:

- Providing traceability of system requirements across the system's life cycle;
- Creating an environment that fosters innovation, experimentation, and demonstration from concept development to fielding;
- Enabling rapid prototyping to deliver capabilities faster and quickly respond to changing threats and requirements;
- Facilitating collaboration to improve integration of system of systems to meet mission needs; and
- Developing a platform and process to support DAF modernization efforts across multiple functional areas to include agile development and modular open systems architecture (MOSA).[98]

Collectively, the DTO and the Digital Engineering Enterprise Office seek to accelerate digital modernization and transformation across the DAF enterprise.[99]

FINDING 13: MBSE and digital engineering methodologies reduce development risk and improve system design and performance.

RECOMMENDATION 17: The Department of the Air Force Chief Architect's Office and the Department of the Air Force Rapid Capabilities Office should work with the Department of the Air Force's Digital Engineering Enterprise Office to apply model-based systems engineering (MBSE) methods across Advanced Battle Management System engineering and sustainment activities and to enable MBSE to serve as a bridge between operator requirements and development teams.

---

[98] R. Jones, 2019, "USAF Digital Engineering Strategy to Implementation," Office of the Deputy Assistant Secretary of the Air Force for Science, Technology, and Engineering, https://www.nist.gov/system/files/documents/2019/04/05/11-5_jones_usaf_mbse_implementation.pdf, April 5.

[99] For more on the DAF's digital transformation, see Air Force Materiel Command, "Digital Campaign," https://www.afmc.af.mil/digital/, and W. Cooley, 2021, "Digital Campaign Overview," Presentation to the Air Force Studies Board, August 6. See also Air Force Studies Board, 2021, "Digital Strategy for the Department of the Air Force: A Workshop Series," https://www.nationalacademies.org/our-work/digital-strategy-for-the-department-of-the-air-force-a-workshop-series.

M&S and VV&A

Modeling is the process of producing a representation of the construction and working of a system; simulation is the operation of a model of the system.[100] Together, M&S is a tool used in systems engineering to inform decisions. Verification, validation, and accreditation (VV&A) refers to three inter-related but unique processes that collect and evaluate evidence to determine whether a model's or simulation's capabilities, accuracy correctness, and usability are sufficient to support its intended uses.[101] In concert with MBSE, M&S and VV&A provides options for trade-space analysis to inform design decisions and to quantify performance during system development. When overlaid with T&E, M&S and VV&A allow systems engineers the opportunity to evaluate with confidence that design implementation is performing according to expectations.[102]

In support of these processes, the DAF has established the Common Simulation Training Environment (CSTE) at the Air Force Life Cycle Management Center's Architecture and Integration Directorate, Wright-Patterson Air Force Base. The role of the CSTE is to create a collaborative environment where training systems and simulators may be better linked to support operators and warfighters. Additionally, the DAF is working to standardize training platform requirements by establishing in 2016 simulator common architecture requirements and standards (SCARS) that implement a modular open system architecture (MOSA) approach and a set of universal standards for DAF simulators.[103] The intent of these activities is to rapidly update technologies on a continual basis, while supporting ABMS and JADC2.

---

[100] See A. Maria, 1997, "Introduction to Modeling and Simulation," in *Proceedings of the 1997 Winter Simulation Conference*, S. Andradottir, K.J. Healy, D.H. Withers, and B.L. Nelson, eds., pp. 7–13, https://ieeexplore.ieee.org/stamp/stamp.jsp?arnumber=640369.

[101] See DoD (U.S. Department of Defense), 2008, "Department of Defense Standard Practice Document of Verification, Validation, and Accreditation (VV&A) for Models and Simulations," MIL-STD-3022, p. 2, January 28.

[102] MITRE Corporation, "Verification and Validation of Simulation Models," *MITRE Systems Engineering Guide*, https://www.mitre.org/publications/systems-engineering-guide/se-lifecycle-building-blocks/other-se-lifecycle-building-blocks-articles/verification-and-validation-of-simulation-models.

[103] See M. Roaten, 2021, "Air Force Looking to Boost Connectivity for Simulators," *National Defense Magazine*, https://www.nationaldefensemagazine.org/articles/2021/7/22/air-force-looking-to-boost-connectivity-for-simulators, July 22.

Digital Twin

A digital twin is a "virtual representation of an object or system that spans its life cycle, is updated from real-time data, and uses simulation, ML, and reasoning to help decision-making."[104] Classes of models include the system model, product model, and process model. A digital twin and digital thread connect these models together in a model-based environment.[105] As an engineering tool, a digital twin provides insights regarding how a system or connected sensors will perform in the future.

Within the DAF, several programs and systems are already adopting digital twins in their engineering plans. For example, the PATS family of systems is using a digital twin for early system modeling to reduce risk and as a built model for system integration and operational support.[106] A digital twin was also used in the design and initial testing for the Air Force's new advanced aircraft trainer jet, the eT-7 Red Hawk, where design, construction, and test flights were all conducted before the first trainer jet was even manufactured and delivered.[107] Similarly, digital twins were used in lieu of flight tests to evaluate competing proposals for the B-52 engine replacement.[108] The DAF also announced plans to create a digital twin for the F-16 to improve sustainment and modernization of its current fleet.[109] The goal

---

[104] M.M. Armstrong, 2020, "Cheat Sheet: What Is Digital Twin?" *IBM Business Operations Blog*, https://www.ibm.com/blogs/internet-of-things/iot-cheat-sheet-digital-twin/, December 4. See also M. Grieves and J. Vickers, 2017, "Digital Twin: Mitigating Unpredictable, Undesirable Emergent Behavior in Complex Systems," pp. 85–113 in *Transdisciplinary Perspectives on Complex Systems*, (F-J. Kahlen, S. Flumerfelt, and A. Alves, eds.), Springer, Switzerland, https://link.springer.com/book/10.1007/978-3-319-38756-7.

[105] NASEM (National Academies of Sciences, Engineering, and Medicine), 2021, *Adapting to Shorter Time Cycles in the United States Air Force: Proceedings of a Workshop Series*, p. 18, The National Academies Press, Washington, DC. Digital thread is defined as "an integrated information flow that connects all the phases of the product life cycle using an accepted authoritative data source." NASEM, 2021, "Adapting to Shorter Time Cycles in the United States Air Force: Proceedings of a Workshop Series," p. 18.

[106] See T. Hitchens, 2021, "Space Force Digital Vision Focuses on Speedy Decisions," *Breaking Defense*, https://breakingdefense.com/2021/05/space-force-digital-vision-focuses-on-speedy-decisions/, May 6, and S. Erwin, 2020, "Space Force Developing a Digital Strategy for Designing and Producing Future Satellites," *Space News*, https://spacenews.com/space-force-developing-a-digital-strategy-for-designing-and-producing-future-satellites/, October 21.

[107] R.S. Cohen, 2020, "Air Force Introduces e-Planes for the Digital Era," *Air Force Magazine*, https://www.airforcemag.com/air-force-introduces-e-planes-for-the-digital-era/, September 14.

[108] S. Waterman, 2020, "Digital Twins Proliferate as Smart Way to Test Tech," *Air Force Magazine*, https://www.airforcemag.com/digital-twins-proliferate-as-smart-way-to-test-tech/, March 15.

[109] B. Brackens, 2021, "Air Force to Develop F-16 'Digital Twin,'" *Air Force Life Cycle Management Center News*, https://www.aflcmc.af.mil/News/Article-Display/Article/2677215/air-force-to-develop-f-16-digital-twin/, June 30.

is to create a "full scale 3D model of the aircraft ... to help address future parts obsolescence, and mitigate supply chain risks ... [to reduce reliance] on legacy manufacturing sources and processes."[110]

For ABMS, the Air Force Research Laboratory's (AFRL's) Munitions Directorate is using digital twins aided by high-performance computing systems and AI/ML in the Digital Enterprise WeaponONE (W1) program to collect data from weapons in-flight, combine that information with data fed from the battlefield, and transmit the aggregated data seamlessly through ABMS back to the digital twin.[111] The goal is to improve accuracy and utility of the digital models. AFRL demonstrated this capability in January 2021 using the Gray Wolf prototype. Gray Wolf is an experimental cruise missile intended to provide clustered deployment against enemy air defenses.[112] During the exercise, Gray Wolf executed a 24-hour air tasking order that enabled missiles to collaborate. W1 collected in-flight data and cross referenced it with information about the battlefield, then used ABMS network to securely transport the information back to the digital twin for analysis. "The all-encompassing, digital, agile, open ecosystem program unites best practices and standards from across government, industry, and academia and applies them to weapons development."[113] Going forward, the W1 program will further advance its digital twin prototypes to enable bi-directional data exchanges with their physical counterparts.

To encourage competition in leveraging even greater use of digital twins, the AFRL is building an online colosseum where commercial vendors' systems are able to compete. "Each vendor could submit a digital twin of a proposed weapons platform for evaluation in 'a kind of Gladiator showdown ... across that particular technology area,' "[114] according to the head of AFRL's Munitions Directorate. Vendors are directed to build their digital twins using the Government's Reference

---

[110] Ibid.

[111] See Air Force Research Laboratory Public Affairs, 2021, "WeaponONE Demonstrates Digital Twin Technologies That Deliver Software-Defined Weapon Capabilities to the Battlefield," *Air Force Life Cycle Management Center News*, https://www.aflcmc.af.mil/News/Article-Display/Article/2478391/weaponone-demonstrates-digital-twin-technologies-that-deliver-software-defined/, January 21.

[112] T. Hitchens, 2021, "AFRL's WeaponONE Aims to Rapidly Build Digital Design, Engineering Tools," *Breaking Defense*, https://breakingdefense.com/2021/02/afrls-weaponone-aims-to-rapidly-build-digital-design-engineering-tools/, February 5.

[113] T. Hitchens, 2021, "AFRL's WeaponONE Aims to Rapidly Build Digital Design, Engineering Tools," *Breaking Defense*, https://breakingdefense.com/2021/02/afrls-weaponone-aims-to-rapidly-build-digital-design-engineering-tools/, February 5.

[114] S. Waterman, 2021, "Air Force Goes All in on Digital Twinning—for Bombs as Well as Planes," *Air Force Magazine*, https://www.airforcemag.com/air-force-goes-all-in-on-digital-twinning-for-bombs-as-well-as-planes/, March 26.

Architecture that provides standards and defines the interfaces the digital model needs to use.

FINDING 14: Significant initiatives are under way across the DAF to encourage the use of digital twins for both legacy platforms and emerging systems.

RECOMMENDATION 18: Building on existing activities in digital engineering and modeling and simulations, the Department of the Air Force Chief Architect's Office and the Department of the Air Force Rapid Capabilities Office should expand the use of digital twins in Advanced Battle Management System development, particularly as new capabilities and technologies are introduced.

COMMON MISSION COMMAND CENTER

An example of a complex architecture and systems integration has been the Family of Systems (FOS) that was initiated in 2010 by the former Under Secretary of Defense for Acquisition, Technology, and Logistics, the former Principal Deputy Under Secretary of Defense for Acquisition, Technology, and Logistics, and the former Air Force SAE. Pulling together a large number of existing and newly developing programs in the FOS, five major elements resulted in a fusion of new technologies and synthesis of major program elements.

That proven process led to a recognition that there is a strong requirement for integrating command, control, communications, and intelligence (C3I) to different program components. Resulting from this need, a government and industry team under the auspices of the DAF RCO has successfully established a prototype capability to support ABMS elements. This prototype, the Common Mission Control Center (CMCC), is a software, hardware, and human machine interface that directs, tasks, and combines multiple missions, to include a large number of weapon systems within a complex C2 framework. The focus is on providing interoperability, mission management, planning and tasking, data fusion from multiple information sources, geo-location status and situational awareness, product management and dissemination, and machine-to-machine data exchange in a secure C2 environment.[115] It has been developed with a well-established open architecture framework using open mission systems/universal command and control interface (OMS/UCI) applications. Additionally, the DAF RCO has led a major effort in protected com-

---

[115] P. Host, 2016, "Air Force, Contractors Working Together on Common Mission Control Center," *Defense Daily*, https://www.defensedaily.com/air-force-contractors-working-together-on-common-mission-control-center/air-force/, June 21.

munications and indications and warnings, which would enhance the multi-layer security of the CMCC.

The 2019–2020 Defense Science Board study on 21st Century Multi-Domain Effects recommended the establishment of an open architecture base in a Joint C3I system that would combine command and control functionalities across the military Services to support complex operations across all domains.[116] The CMCC provides this type of integrated command and control construct and may be scalable to support ABMS functions and JADC2 mission requirements going forward. If CMCC serves as phase zero for ABMS, then future upgrades would be refreshed every year or two to fully integrate with new capability releases in a disciplined and symbiotic way. The capability upgrades would be continuous to prevent obsolescence and to ensure that ABMS is equipped to address evolving adversarial threats.

RECOMMENDATION 19: The Department of the Air Force Rapid Capabilities Office should consider scaling the Common Mission Control Center and designate it as phase zero for the Advanced Battle Management System.

---

[116] DoD, 2020, "21st Century Multi-Domain Effects Executive Summary," Defense Science Board, https://dsb.cto.mil/reports/2020s/FINALMDEExecutiveSummary.pdf, September.

3

Governance

> Stove-piped, single-service solutions that don't integrate for joint force commanders are of little use in future joint warfare.
> —General David W. Allvin, Vice Chief of Staff, U.S. Air Force[1]

Governance of the Advanced Battle Management System (ABMS) requires both a command structure and decision structure. Command structure determines organization hierarchy and interrelationships across organizations, whereas decision structure focuses on decision-making and execution. As a multi-platform, multi-system construct, the command structure of ABMS falls under the Office of the Assistant Secretary of the Air Force for Acquisition, Technology, and Logistics (SAF/AQ). When Department of the Air Force (DAF) leaders in 2019 reintroduced ABMS as an integrated system of systems in support of the Joint All-Domain Command and Control (JADC2) framework, they selected a Chief Architect to "create and manage family of systems trade space, design margins, and define interfaces and standards to ensure interoperability across domains and permissive to highly contested environments."[2] He was also tasked with coordinating the disparate

---

[1] D. Allvin, 2021, "Why We Need the Advanced Battle Management System," *DefenseOne*, https://www.defenseone.com/ideas/2021/05/why-we-need-advanced-battle-management-system/173861/, May 6.

[2] A. McCullough, 2019, "ABMS Expected to Pick Up Speed with New Chief Architect in Place," *Air Force Magazine*, https://www.airforcemag.com/abms-expected-to-pick-up-speed-with-new-chief-architect-in-place/, March 10.

activities of individual programs (that feed into ABMS) led by program managers with their own funding and performance schedules.

In November 2020, the former SAF/AQ directed that ABMS management be transferred from the Chief Architect's Office to the Department of the Air Force Rapid Capabilities Office (DAF RCO) as the integrating Program Executive Office (PEO). "Warfighters are now ready to field and operationalize specific ABMS capabilities across their mission areas. Consequently, ABMS is now graduating into a steady-state demonstration-deployment phase."[3]

Under this new command structure, the DAF RCO, tasked as the ABMS PEO, is responsible for:

- Drafting the ABMS acquisition strategy and subsequent changes in coordination with the Chief Architect;
- Accomplishing a comprehensive business review conducted by the Air Force Audit Agency that will inform the ABMS acquisition strategy;
- Drafting overarching ABMS architectures and standards for the Chief Architect's approval, while the ABMS PEO will have approval authority for all lower-level standards not at the system level;
- Chairing all design reviews below the ABMS architecture review board (ARB);
- Delivering and integrating all ABMS capabilities for inclusion in architecture evaluation on-ramps; and
- Executing the ABMS program according to the approved ABMS acquisition strategy and ARB decisions.

The Chief Architect will:

- Codify ABMS technical requirements derived from the Air Force and Space Force Service Chief-approved requirements documents and on-ramp results;
- Facilitate an integrating enterprise digital architecture and standards across the DAF, Combatant Commands, partnering Services, agencies, and other mission partners;
- Chair the ABMS ARB between on-ramps;
- Provide inputs to the ABMS acquisition strategy;
- Engage with both DAF senior stakeholders and external senior stakeholders to ensure unity of effort and division of engagement responsibilities; and

---

[3] W. Roper, 2020, "Advanced Battle Management System Management Construct," Memorandum for Record, https://insidedefense.com/sites/insidedefense.com/files/documents/2020/nov/11242020_abms.pdf, November 24.

- Establish and provide model-based systems engineering (MBSE) and other collaboration tools across the DAF to enable digital engineering.

The Service Acquisition Executive (SAE) will:

- Retain decision authority for all aspects of ABMS to include approving both the ABMS technical architecture and acquisition strategy and all subsequent changes; and
- Resolve differences between the Chief Architect, ABMS PEO, and related PEOs.[4]

The committee supports this governance structure and considers it a positive progression consistent with the evolving nature of a complex system like ABMS. The DAF RCO has a solid record for developing, acquiring, and fielding critical combat capabilities through the use of commercial technologies and equipment, defense-wide technology development efforts, and accelerated acquisition methods to counter the increasing pace of the threat evolution. In its nearly 20-year history, the DAF RCO has successfully developed sophisticated and advanced weapons systems to include the X-37B orbital test vehicle, B-21 Raider long-range strike bomber, an unmanned space test platform for the U.S. Space Force, a surface-to-air missile system, and other highly classified systems.[5]

Beyond the roles and responsibilities detailed by the SAF/AQ, it is important to note that the command structure to maintain and sustain ABMS also needs to be defined and established before the end of the initial deployment. Roles, responsibilities, and funding schemes for maintenance and sustainment need to be defined for the system to thrive past initial deployment.

From a decision-making structure, ABMS requires more than just internal-DAF coordination and approvals. As a contributor to the JADC2 framework, ABMS requires inter-Service and multi-national coordination with America's partners and allies guided by a set of mutually agreed upon operating standards and policies. This will require a U.S. Department of Defense (DoD)-level governance structure with true decision-making authorities. The current JADC2 cross functional team (CFT) led by the J6 includes too many participants and is not sufficiently empowered to make needed high-level decisions. Instead, a higher-level, joint decision-making body needs to be established to provide cross-Service decisions regarding command authorities for all domain operations, human-machine decisions, interoperability, and shared technologies. The challenge that JADC2 presents is that each

---

[4] W. Roper, 2020, "Advanced Battle Management System Management Construct."

[5] See U.S. Air Force, 2020, "Rapid Capabilities Office Fact Sheet," https://www.af.mil/About-Us/Fact-Sheets/Display/Article/2424302/rapid-capabilities-office/, November 23.

Service, combatant command, and DoD agency is developing its own command and control (C2) system with minimal coordination and deconfliction. The end result is a multitude of disconnected, stove-piped networks that may not interoperate in a multi-domain environment.

Governance during operations is also of concern. While the new Joint Warfighting Concept (JWC) has been developed, it remains unclear how data flows to desired actions will be prioritized using JADC2 supported systems. For example, a central thrust of JADC2 is to compress the observe-orient-decide-act (OODA) loop by optimizing the flow of data. However, in the context of a global positioning system (GPS)-denied, electronic warfare (EW), or cyber-compromised environments, it is unclear how decision-making will be conducted, particularly if data is transported through automation.

Moreover, tactical level integration requires that all-domain operations continue after communications with the joint headquarters have been denied. Distributed units must possess both the understanding and authority to act under general commander's intent in the absence of more-specific command orders. This will require significant rethinking of the distribution and assignment of authorities, particularly when operational decisions have the potential to escalate conflicts between nuclear powers. More importantly, considerations of trust must be evaluated and balanced against risks: Where and how will the military rely on and accept the information and abilities of unknown or new agents, especially when lives and major assets are at stake? Can lower echelons be entrusted to make strategic-level decisions? These issues are further complicated when multi-national partners are factored into the decision-making space.[6]

Another issue of concern is that each military Service is developing and selecting C2 solutions outside their domain of control with the intent of resolving joint mission requirements. Already, Army leaders have expressed concerns that ground troops cannot adopt an air-centric command system for future all-domain operations. According to the former Deputy Commanding General of the Army Futures Command and the Director of the Futures and Concepts Center, "ABMS cannot be the sole solution, because it doesn't account for, in some cases, the scale or the unique requirements of all the other services. ... Army scaling issues have to be considered in any kind of framework that's put together in the future. Other services might be looking at the scale of hundreds, where the Army is looking at

---

[6] See W. Perkins and A. Olivieri, 2018, "On Multi-Domain Operations: Is NATO Today Sufficiently 'Joint' to Begin Discussions Regarding Multi-Domain Command and Control?" *The Journal of the (JPACC) Joint Air Power Competence Centre*, (26):16–23, https://www.japcc.org/on-multi-domain-operations/.

a scale of thousands."[7] The committee sees the resolution of these issues as central to ABMS and the larger JADC2 framework.

FINDING 15: The current JADC2 CFT led by the J6 is a positive first step, but it includes too many participants and is not sufficiently empowered to make needed high-level decisions.

RECOMMENDATION 20: The Joint Staff J6 or a designated U.S. Department of Defense executive agent should establish an authoritative Joint-level body to address and resolve technical, operational, and command decisions for all contributors to the Joint All-Domain Command and Control framework.

ORGANIZATION INTEGRATION

Beyond the technical challenges of establishing a joint C2 environment, integrating the wide-ranging ABMS ecosystem within JADC2 will require both organizational and human considerations. From an organizational perspective, ABMS requires the ability to work across military Services, defense agencies, and multi-national partners, each with its own distinct culture and operating norms. Incompatibility between organizations will require the commitment of the military Services—both individually and collectively—to resolve. To break down these vertical silos and achieve meaningful and effective joint interoperability at all levels, from tactical to strategic, DoD needs to create a unified vision supported by common tactics, techniques, and procedures (TTPs). The J6 CFT is positive first step in advancing a common understanding of joint interoperability and setting universal standards, but more needs to be accomplished. This is an area where industry practices may provide a useful guide.

Large organizations tend to be characterized by autonomous units that are either unwilling or unable to coordinate and integrate with other units.[8] Individuals within a division tend to interact more within their own units than with outside groups. This results in fragmentation, division, and disconnection within the broader organization—in essence, creating organizational silos. There are three

---

[7] S.J. Freedberg, Jr., 2020, "ABMS Can't Be 'Sole Solution' for Joint C2, Army Tells Air Force—Exclusive," *Breaking Defense*, https://breakingdefense.com/2020/01/abms-cant-be-sole-joint-c2-solution-army-tells-air-force-exclusive/, January 22.

[8] See A.C. Edmondson, S. Jang, and T. Casciaro, 2019, "Cross-Silo Leadership," *Harvard Business Review*, https://hbr.org/2019/05/cross-silo-leadership, May-June, and S. Billingsley, 2021, "Organizational Silos," *LinkedIn*, https://www.linkedin.com/pulse/organizational-silos-scott-billingsley/, May 27.

common influences that result in silos: internal, organizational, and external, which interact to further reinforce the strength of the silos.[9]

Within the DoD community, internal influences are characterized by each military Service's and agency's structure and culture; organizational influences are the TTPs and operating norms that are specific to the program unit; and the external environment comprise of requirements from combatant commanders, multi-national partners, other military Services, federal agencies, among others. Each influence imposes on the core unit that ultimately reinforces an insular, parochial, and stove-piped structure. When overlaid with a complex framework like JADC2, the process of overcoming Service-centric silos becomes more challenging.

In order to achieve an effective and interoperable (not just complementary) joint C4 enterprise architecture, organizational and cultural barriers need to be lowered through horizontal integration. The challenge of interconnecting cross-Service networks may be resolved through technological advancements, but the challenge of interconnecting cross-Service organizations requires social integration to develop cooperative partnerships and trust. Effective horizontal integration requires leaders to "connect the [organization's] knowledge bases, build social relationships among people and shape a shared sense of identity, all supported by a standardized technological infrastructure."[10] This may be accomplished through four areas of action:

- Operational integration through standardization of the technological infrastructure;
- Intellectual integration through the development of a shared knowledge base;
- Social integration through collective bonds for performance; and
- Emotional integration through the creation of a common identity and purpose.[11]

Figure 3.1 provides a framework for considering organizational integration. To further decompose vertical silos, industry employs six basic steps:

---

[9] Select Strategy, LLC, 2002, "Improving Performance by Breaking Down Silos: Understanding Organizational Barriers," https://selectstrategy.com/download/Breaking%20down%20organizational%20barriers.pdf.

[10] S. Ghoshal and L. Gratton, 2002, "Integrating the Enterprise," *MIT Sloan Management Review*, https://sloanreview.mit.edu/article/integrating-the-enterprise/, October 15.

[11] S. Ghoshal and L. Gratton, 2002, "Integrating the Enterprise," *MIT Sloan Management Review*, https://sloanreview.mit.edu/article/integrating-the-enterprise/, October 15, p. 33.

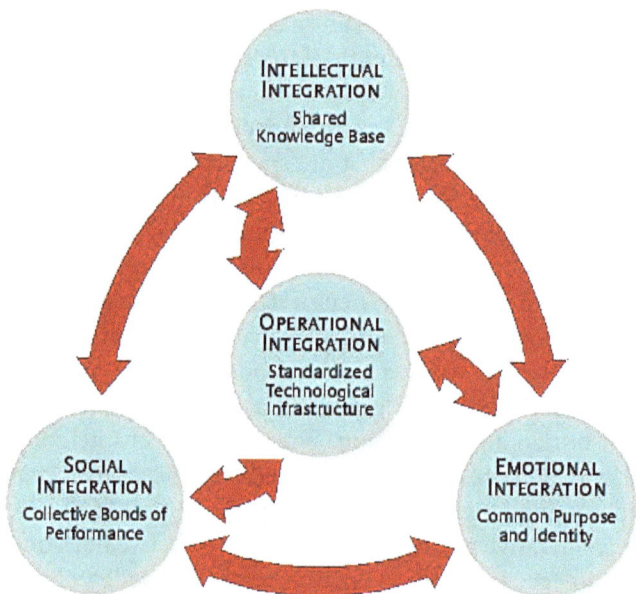

FIGURE 3.1 Organization integration framework. SOURCE: S. Ghoshal and L. Gratton, 2002, "Integrating the Enterprise," MIT Sloan Management Review, https://sloanreview.mit.edu/article/integrating-the-enterprise.

1. Communicate a unified vision;
2. Create shared accountabilities;
3. Bring teams together;
4. Get leaders on board;
5. Incorporate collaboration tools; and
6. Shift mindsets and behavior with training.[12]

For ABMS and other contributors to JADC2, the shared mission to sustain the Joint Force's military advantages by helping decision makers to act on information well inside the adversaries' OODA loop provides a unifying vision upon which to act. Leaders across all of DoD are fully supportive of JADC2 and have worked jointly to advance the concept through inter-Service agreements and ex-

---

[12] I. Cornett, 2018, "6 Strategies for Breaking Down Silos in Your Organization," *Eagle's Flight*, https://www.eaglesflight.com/blog/6-strategies-for-breaking-down-silos-in-your-organization, October 25.

perimentations.[13] For example, the Air Force and Naval Studies Boards at the National Academies of Sciences, Engineering, and Medicine are hosting an inter-Service meeting between the Air Force and Navy to discuss their contributions to JADC2. The meeting builds on the ongoing partnership between the DAF RCO and the Naval Information Warfare Systems Command (NAVWAR) to establish integrated approaches between ABMS and Project Overmatch.[14] The J6's leadership on JADC2 and the participation of various stakeholders in the CFT encourages even greater collaboration and shared accountabilities. What is missing, however, are shifting individual mindsets and behavior with training and incentives to forego entrenched organizational cultures and control for the wider good of the joint and multi-national defense ecosystem.

> **FINDING 16:** ABMS and JADC2 will require both horizontal and vertical integration across capabilities, functions, organizations, command echelons, among others. This is more than just a technical problem to be solved by technical systems but includes aspects of workforce-system integration, training, trust, ownership, and control as well as underlying social aspects that have yet to be addressed.

> **FINDING 17:** Integrating the JADC2 enterprise will be a continuous and evolving process. This process will require two parallel streams. The first will be the system evolution that addresses changes in technology, the environment, emerging threats, mission requirements, and relevant tools needed to support the second stream: evolution of the TTPs and related organizational, social, and emotional challenges.

> **RECOMMENDATION 21: The Joint Chiefs and military department secretaries should tackle the cultural, social, and emotional barriers to true Joint Warfighting Concept (JWC) horizontal integration if the Advanced Battle Management System and the larger Joint All-Domain Command and Con-

---

[13] See for example A. Eversden, 2020, "US Army, Air Force Sign Agreement to Develop Joint All-Domain Concept," *C4ISRNet*, https://www.c4isrnet.com/battlefield-tech/it-networks/2020/10/05/us-army-air-force-sign-agreement-to-develop-joint-all-domain-concept/, October 5, J. Koester, 2020, "JADC2 'Experiment 2' Provides Looking Glass into Future Experimentation," *Army News*, https://www.army.mil/article/234900/jadc2_experiment_2_provides_looking_glass_into_future_experimentation, April 23, and T. Hitchens, 2020, "Air Force Chief Seeks Navy Chief's Cooperation on JADC2," *Breaking Defense*, https://breakingdefense.com/2020/10/air-force-chief-seeks-navy-chiefs-cooperation-on-jadc2/, October 21.

[14] D.W. Small, 2021, "Project Overmatch," Presentation to the Air Force ABMS Committee, March 3, and R. Walden, 2021, "ABMS Perspectives from the Air Force Rapid Capabilities Office," Presentation to the Air Force ABMS Committee, January 22.

trol constructs are to enable the truly joint and multi-national integrated operations envisioned by the JWC.

HUMAN FACTORS

While ABMS is intended to deliver the "decision advantage," humans remain central to decision-making. According to the Vice Chief of the Air Force, "ABMS [is] a way for humans and algorithms to manage mass quantities of data securely from multiple sources through multiple domains that is ingested, fused, processed, and presented in a manner useful to commanders."[15] For highly strategic decisions such as NC3, humans are even more vital and cannot easily be replaced by machines.[16]

Evolving the current technical and operational environment to fulfill the vision of all-domain operations will require significant changes at both the human and technical levels. The extent of the changes in just one of these dimensions—let alone on both dimensions concurrently—will require increased and expanded levels of trust and verification. Examples of broadened activities requiring expanded levels of trust include sensors and shooter access and prioritized use of scarce communications, spectrum, and computing resources. Coupled with advancements in AI and technology that allow humans to take advantage of operations at machine speed, there are multiple implications for ABMS and future military operations.

Human Systems Integration

As highlighted by the National Security Commission on Artificial Intelligence (NSCAI), "AI cannot magically solve problems.... Harnessing data, hardening and packaging laboratory algorithms so they are ready for use in the field, and adapting AI software to legacy equipment and rigid organizations all require time, effort, and patience. Integrating AI often necessitates overcoming substantial organizational and cultural barriers, and it demands top-down leadership."[17] To integrate humans with AI/machine learning (ML) in the context of ABMS and JADC2, considerations must be given to training, tools and methodologies, system operations, occupational health and safety, and ethics.[18]

According to the Under Secretary of Defense for Research and Engineering,

---

[15] D. Allvin, 2021, "Why We Need the Advanced Battle Management System."

[16] See R.K.C. Hersman, E. Brewer, and S. Claeys, 2020, "NC3 Challenges Facing the Future System," Center for Strategic and International Studies, https://csis-website-prod.s3.amazonaws.com/s3fs-public/publication/207009_NC3_Challenges_Facing_Future_System_v7.pdf, July.

[17] NSCAI, 2021, "National Security Commission on Artificial Intelligence," p. 21.

[18] See National Research Council, 2007, *Human-System Integration in the System Development Process: A New Look*, The National Academies Press, Washington, DC.

Human Systems Integration (HSI) is the comprehensive, interdisciplinary management and technical approach applied to system development and integration as part of a wider systems engineering process to ensure that human performance is optimized to increase total system performance and minimize total system ownership costs. HSI enables the systems engineering process and program management effort that provides integrated and comprehensive analysis, design, and assessment of requirements, concepts, and resources for seven domains: human factors engineering (HFE), manpower, personnel, training, safety and occupational health (SOH), force protection and survivability, and habitability.[19]

Humans are included in their roles as operators, designers, maintainers, engineers, while systems include hardware, software, and design, acquisition, security, and other processes.[20]

The dual goals of HSI are to (1) achieve human performance effectiveness during all stages of the system's life cycle to include testing, operation, maintenance, support, transport, demilitarization, and disposal; and (2) ensure overall human performance possesses the necessary knowledge and competencies to support mission tasking.[21] As a management framework, HSI facilitates trade-offs among its seven domains and other systems engineering domains, but does not replace individual domain activities, responsibilities, or reporting channels.[22] More importantly, HSI enables the collection of quantifiable and measurable impacts to overall system design.[23]

Within ABMS, the complexities of integrating almost 30 different product lines with operators, engineers, developers, testers, trainers, and others is a daunting challenge. "There's so many people in between information, moving between different nodes in the decision chain ... the idea with ABMS is that the people are no longer the glue. The information flows everywhere all at once. The people are the assessors, the analyzers, the feedback providers that help the analytics ... to be better and better."[24] HSI can thus provide a robust framework for ensuring that

---

[19] DDR&E (U.S. Department of Defense Research and Engineering Enterprise), "Human Systems Integration," https://ac.cto.mil/hsi/, accessed September 18, 2021.

[20] DAU (Defense Acquisition University), "Human Systems Integration," https://www.dau.edu/acquipedia/pages/articledetails.aspx#!489, accessed September 18, 2021.

[21] Ibid.

[22] DDR&E, "Human Systems Integration."

[23] See USAF Directorate of Human Performance Integration, "Air Force Human Systems Integration Handbook," https://www.acqnotes.com/Attachments/Air%20Force%20Human%20System%20Integration%20Handbook.pdf, and USAF Human Systems Integration Office, 2009, "Human Systems Integration Requirements Pocket Guide," September.

[24] S. Maucione, 2020, "Air Force Using Agile Approach to Connect Systems for Battle," *Federal News Network*, https://federalnewsnetwork.com/air-force/2020/01/air-force-using-agile-approach-to-connect-systems-for-battle/, January 21.

ABMS design and development can effectively integrate human capabilities and limitations.

> **RECOMMENDATION 22: The Department of the Air Force Rapid Capabilities Office should incorporate human systems integration methodologies into the Advanced Battle Management System to ensure that all human users are fully and effectively integrated with current and future systems elements.**

Training, Culture, and Other Considerations

ABMS is more than just a technical problem to be solved by AI and advanced technologies. It requires training and an emphasis on culture and awareness, particularly because it involves working with sister Services and multi-national partners and allies. As previously discussed, organizational silos across the DoD limit effective communication, collaboration, and interoperability. However, this challenge is further compounded by rigid boundaries between select warfighting communities, because the TTPs within these communities are so distinct. When coupled with participation from multi-national partners and allies, cultural, geographical, and language barriers exacerbate the difficulties of achieving effective coordination, compatibility, and ultimately, interoperability.

Another area that demands further attention is ethical use in AI. The DoD introduced in 2020 five guiding principles for ethical development of AI capabilities:

- *Responsible:* DoD personnel will exercise appropriate levels of judgment and care while remaining responsible for the development, deployment, and use of AI capabilities;
- *Equitable:* The department will take deliberate steps to minimize unintended biases in AI capabilities;
- *Traceable:* The department's AI capabilities will be developed and deployed such that relevant personnel possess an appropriate understanding of the technology, development, processes, and operational methods applicable to AI capabilities, including with transparent and auditable methodologies, data sources, and design procedures and documentation;
- *Reliable:* The department's AI capabilities will have explicit, well-defined uses, and the safety, security, and effectiveness of such capabilities will be subject to testing and assurance within those defined uses across their entire life cycles; and
- *Governable:* The department will design and engineer AI capabilities to fulfill their intended functions while possessing the ability to detect and

avoid unintended consequences, and the ability to disengage or deactivate deployed systems that demonstrated unintended behavior.[25]

As ABMS seeks to leverage AI to enhance and accelerate decision-making, questions concerning ethical use will likely emerge. The level of human engagement in decision-making must continue to align with the DoD Directive on Autonomy in Weapons Systems[26] and the law of war.[27] Furthermore, the consideration of potential accidents, human-machine errors and miscues, and even sabotage need to be carefully evaluated. The challenge is balancing the need and associated risks for accelerated decision-making with accuracy, reliability, and precision.

One way to ameliorate misguided AI conclusions is to provide more diverse data samples and broader context to the data. Another approach is to use hyper-

---

[25] C.T. Lopez, 2020, "DoD Adopts 5 Principles of Artificial Intelligence Ethics," *DoD News*, https://www.defense.gov/News/News-Stories/Article/Article/2094085/dod-adopts-5-principles-of-artificial-intelligence-ethics/, February 25.

[26] Deputy Secretary of Defense, 2012, "Autonomy in Weapon Systems," DoD Directive 3000.09, Incorporating Change 1, May 8, 2017, pp. 2–3, https://www.esd.whs.mil/portals/54/documents/dd/issuances/dodd/300009p.pdf, November 21. This directive states, "It is DoD policy that … autonomous and semi-autonomous weapon systems shall be designed to allow commanders and operators to exercise appropriate levels of human judgment over the use of force…. Persons who authorize the use of, direct the use of, or operate autonomous and semi-autonomous weapon systems must do so with appropriate care and in accordance with the law of war, applicable treaties, weapon system safety rules, and applicable rules of engagement (ROE)…. Human-supervised autonomous weapon systems may be used to select and engage targets, with the exception of selecting humans as targets, for local defense to intercept attempted time-critical or saturation attacks for: (a) Static defense of manned installations. (b) Onboard defense of manned platforms…. Autonomous weapon systems may be used to apply non-lethal, non-kinetic force, such as some forms of electronic attack, against materiel targets in accordance with DoD Directive 3000.03E…. Autonomous or semi-autonomous weapon systems intended to be used in a manner that falls outside the policies … must be approved by the Under Secretary of Defense for Policy (USD(P)); the Under Secretary of Defense for Acquisition, Technology, and Logistics (USD(AT&L)); and the CJCS before formal development and again before fielding."

[27] General Counsel of the DoD, 2015, *Department of Defense Law of War Manual* updated December 2016, p. 353, https://ogc.osd.mil/Portals/99/department_of_defense_law_of_war_manual.pdf, June, which states that "Although no law of war rule specifically restricts the use of autonomy in weapon systems, other rules may apply to weapons with autonomous functions. For example, to the extent a weapon system with autonomous functions falls within the definition of a 'mine' in the [Convention against Chemical Weapons] CCW Amended Mines Protocol, it would be regulated as such. In addition, the general rules applicable to all weapons would apply to weapons with autonomous functions. For example, autonomous weapon systems must not be calculated to cause superfluous injury or be inherently indiscriminate."

parameter tuning to refine the algorithm.[28] Cross-validation may also be used to separate data into various partitions and train multiple algorithms on these partitions to improve the soundness of the model.[29] Furthermore, a different algorithm or sets of algorithms may be required to better fit the data set.[30] Last, to keep pace with increased computing power and emerging technologies that complement AI, streamlined acquisition processes and improved TTPs may also be warranted.

In spite of improvements with AI data accuracy, one industry survey on AI and big data found the primary obstacles for large organizations to successfully transition to modern, data-centric environments are cultural rather than technical.[31] For this reason, training and the creation of a cadre of highly qualified experts (HQEs) that together bring a combination of policy, operations, and technical expertise is needed. Areas for specific focus may include the following:

- *AI/ML:* These rapidly developing technologies have a significant role in command, control, and communications (C3) at all levels—strategic, operational, and tactical. Creating capabilities and using them effectively will require a broad range of knowledge and skillsets to include concepts of operations to data management, and from cybersecurity to testing.
- *MBSE:* As mentioned in the section on MBSE in Chapter 2, MBSE provides a robust framework for reducing development risk, improving system performance, institutionalizing rigor and precision into the design process, and enhancing knowledge transfer. This is particularly true for key cross-cutting capabilities, such as operational performance and cybersecurity. With ABMS's reliance on DevSecOps, it will be important to ensure that developers do not make incremental improvements to obsolete and un-evolvable technologies or develop applications that are incompatible with the broader architecture framework. While even the most seasoned engineers cannot be

---

[28] See A. Lee, 2019, "Why You Should Do Feature Engineering First, Hyperparameter Tuning Second as a Data Scientist," *Towards Data Science*, https://towardsdatascience.com/why-you-should-do-feature-engineering-first-hyperparameter-tuning-second-as-a-data-scientist-334be5eb276c, April 21, and Prabhu, 2018, "Understanding Hyperparemeters and Its Optimisation techniques," *Toward Data Science*, https://towardsdatascience.com/understanding-hyperparameters-and-its-optimisation-techniques-f0debba07568, July 3.

[29] See L. Quintanilla, N. Schonning, and N. Kershaw, 2021, "Train a Machine Learning Model Using Cross Validation," Microsoft, https://docs.microsoft.com/en-us/dotnet/machine-learning/how-to-guides/train-machine-learning-model-cross-validation-ml-net, October 5.

[30] See L. Quintanilla, B. Achtman, B. Ozdemir, N. Schonning, Y. Victor, and N. Kershaw, 2021, "How to Choose an ML.NET Algorithm," Microsoft, https://docs.microsoft.com/en-us/dotnet/machine-learning/how-to-choose-an-ml-net-algorithm, March 31.

[31] See New Vantage Partners, 2021, "Big Data and AI Executive Survey 2021," New Vantage Partners, LLC, https://c6abb8db-514c-4f5b-b5a1-fc710f1e464e.filesusr.com/ugd/e5361a_d59b4629443945a0b0661d494abb5233.pdf.

expected to know all about this vast evolving ecosystem, training a cadre of experienced engineers who understands MBSE can provide an essential link between visionary concepts and planned architectures with executable operational and development baselines.
- *Cybersecurity:* ABMS is intended to be an Internet of Things (IoT). Shared access across dispersed networks, platforms, and classifications exposes the ecosystem's vulnerabilities and subject it to potential cyberattacks. Training experts who are knowledgeable in cutting-edge cyber defense software and technologies is thus critical to protecting ABMS from malicious attacks and intrusion.
- *Intelligence:* The threat environment is changing rapidly with the influx of inexpensive and advanced commercial technologies that enable adversaries to adapt quickly to erode the United States' decision superiority. The need for intelligence analysts and assessors (i.e., those who can accurately and credibly assess the threats posed by adversaries and other malicious actors) is vital. Technology in itself cannot solve the security challenge. It may help to shorten the OODA loop cycle, but a lack of understanding regarding true adversarial capabilities and intent will weaken America's stance even further.[32]
- *Red teaming:* ABMS is designed to evolve with the emergence of newer and more advanced technologies and with changes in the threat environment. As such, testing and evaluation methodologies must remain fluid and dynamic to adjust to changes in the technological and security environments.
- *Military operations:* Technical HQEs must be augmented by experts that understand what ABMS and JADC2 is trying to accomplish. These are the warfighters, commanders, operators, and end users who conduct military operations and understand their needs, challenges, functions, and TTPs.
- *Culture:* ABMS and JADC2 are intended to work as a globally integrated force without regard to geographic and organizational boundaries. This requires not only working with 11 combatant commanders, but the participation of multi-national partners and allies, as well. Establishing a cadre of analysts and operators who possess knowledge and understanding regarding the TTPs, cultural etiquette, languages, and norms of ally services and partners will be critical for ensuring seamless and coordinated operations within an all-domain operating environment.

---

[32] See D. Sukman and C. Davis, 2020, "Divided We Fall: How the U.S. Force Is Losing Its Joint Advantage Over China and Russia," *Military Review*, https://www.armyupress.army.mil/Journals/Military-Review/English-Edition-Archives/March-April-2020/Sukman-Divided/, March-April.

FINDING 18: ABMS cannot be resolved by technology alone. Non-materiel aspects of DOTMLPF-P must also be addressed.

RECOMMENDATION 23: The Department of the Air Force's Assistant Secretary of the Air Force for Acquisition, Technology, and Logistics and the Deputy Chief of Staff for Strategy, Integration, and Requirements should consider and weave personnel, cultural, training, and other non-materiel doctrine, organization, training, materiel, leadership, education, personnel, facilities, and policy issues into designs and implementation plans for the broader Advanced Battle Management System ecosystem.

FINDING 19: The broad applications of automation and AI envisioned in ABMS (and JADC2) raise ethical risks and considerations given they involve removing humans from phases in the kill chains and other significant operations.

RECOMMENDATION 24: Civilian and military leaders in the Department of the Air Force, Joint Staff, and the Office of the Secretary of Defense should ensure that the ethical use of artificial intelligence is examined and addressed in the Advanced Battle Management System's (and in other systems supporting the broader Joint All-Domain Command and Control framework's) design, operation, staffing, and training, as dictated by policy and the law of war.

RECOMMENDATION 25: The Air Education Training Command should establish a curriculum that would train or recruit highly qualified experts in artificial intelligence/machine learning, model-based systems engineering, cybersecurity, intelligence assessment, and test and evaluation for information technology, software, and hardware who can work with experts in military operations and culture.

4

Challenges and Opportunities

As an evolving family of systems construct, the Advanced Battle Management System (ABMS) requires both technical and non-technical solutions. While significant progress has been made to transition ABMS from on-ramp demonstrations and experimentations to fielding specified solutions through capability releases, the evolving nature of the threat environment necessitates continuous security enhancements, software and hardware refreshes and upgrades, and the introduction of newer and more advanced technologies. As such, the committee is able to provide only an assessment of ABMS technology and planned system integration architecture as it exists during the committee's review. The committee has identified two high-level areas for further consideration. Additional insights are provided in the accompanying classified annex.

INTEROPERABILITY

There are many barriers—technical, organizational, cultural, and procedural—to achieving universal interoperability, which have been discussed in preceding chapters. One of the greatest technical hurdles is linking all systems to all domains and ensuring their interoperability. A 2015 study found that interconnected systems are subject to the CACE principle: changing anything changes everything.[1] "Op-

---

[1] D. Sculley, G. Holt, D. Golovin, E. Davydov, T. Phillips, D. Ebner, V. Chaudhary, and M. Young, 2015, "Machine Learning: The High-Interest Credit Card of Technical Debt," *Advanced in Neural Information Processing Systems* 28 (NIPS 2015).

erational reliance on the combination of separate systems increases vulnerability to emergent effects. It creates a strong entanglement: improving an individual component model may actually make the system accuracy worse if the remaining errors are more strongly correlated with the other components."[2]

Moreover, joint integration does not equate to interoperability, and past evidence has shown that joint integration only results in deconfliction or synergy.[3] For ABMS and other contributors to the Joint All-Domain Command and Control (JADC2) framework, developing an elegant solution that would integrate all systems across all domains remains challenging.

To achieve JADC2-level interoperability, the U.S. Department of Defense (DoD) should consider the following:

- Establish a DoD executive agent (EA) or joint program office (JPO) to set common operational and data standards. This does not mean that all systems need to be built to the same technical specifications. However, it does require the systems to have the ability to interoperate outside of their Service-centric domains. The EA or JPO would provide guidance and oversight for the systems in development; prioritize requirements in concert with the Joint Requirements Oversight Council (JROC); and promote the use of open architecture standards.
- Promote the use of open system architecture (OSA) to facilitate modularity and interoperability among the systems. To achieve cross-Service and multi-domain compatibility, the DoD EA or JPO needs to identify an adaptable and customizable OSA that can be tailored specifically to the Service's unique requirements, while also permitting the system to interoperate with other systems in the JADC2 framework.
- Promote the use of model-based systems engineering (MBSE) to reduce development risk and improve system performance across all systems.
- Partner with industry and other government agencies to adopt best practices, particularly from organizations that have successfully executed large-scale, enterprise-wide digital transformation. This may include becoming members of industry associations, such as the Institute of Electrical and Electronics Engineers (IEEE), the Consultative Committee for Space Data Systems (CCSDS), Object Management Group (OMG), among others.

---

[2] R. Danzig, 2018, "Technology Roulette: Managing Loss of Control as Many Militaries Pursue Technological Superiority," Center for New American Security, https://s3.us-east-1.amazonaws.com/files.cnas.org/documents/CNASReport-Technology-Roulette-DoSproof2v2.pdf?mtime=201806280772101&focal=none, June.

[3] See W.O. Odom and C.D. Hayes, 2014, "Cross-Domain Synergy: Advancing Jointness," *Joint Forces Quarterly* 73(2nd quarter):123–128.

- Coordinate with multi-national partners and allies to test their systems to ensure that they are complementary with the U.S. military's ecosystem.

INTELLIGENCE

The adversaries of today and the future are highly sophisticated, diverse, and unpredictable. The range of military missions and operations that the Department of the Air Force (DAF) will have to undertake will require a broad spectrum of capabilities from tactical to strategic. "This increasingly complex security environment is defined by rapid technological change, challenges from adversaries in every operating domain, and the impact on current readiness from the longest continuous stretch of armed conflict in our Nation's history.... These changes require a clear-eyed appraisal of the threats we face, acknowledgement of the changing character of warfare, and a transformation of how the Department conducts business."[4]

As mentioned in the preceding chapter, technology in itself is insufficient for addressing challenges to the nation's security. Technology, coupled with an understanding of the threats imposed, will enable improvements and shorten the timeline in the DAF's observe-orient-decide-act (OODA) loop cycle. To achieve a realistic assessment of adversarial capabilities and intent, the DAF should consider the following:

- Establish a DAF net assessment capability (similar to OSD's Office of Net Assessment) that could identify emerging trends, threats, and opportunities; red team and conduct wargames to test DAF capabilities; and provide independent research and analyses that leverage latest thinking and relevant historical lessons to better understand the adversaries' doctrines, operational concepts, and technical capabilities.[5]
- Work with the U.S. Strategic Command and other Combatant Commands to connect ABMS mission requirements, to include nuclear command, control, and communications (NC3), with the threat environment, and enhance the ecosystem's capabilities as the threat environment evolves.

---

[4] DoD, 2018, *Summary of the 2018 National Defense*, https://dod.defense.gov/Portals/1/Documents/pubs/2018-National-Defense-Strategy-Summary.pdf, p. 2.

[5] See Office of the Chief Management Officer of the Department of Defense, 2020, "DoD Directive 5111.11: Director of Net Assessment," https://www.esd.whs.mil/Portals/54/Documents/DD/issuances/dodd/511111p.pdf, April 14.

MAJOR RECOMMENDATIONS

To summarize, the committee has categorized its recommendations into two broad categories: technical and non-technical.

Technical

RECOMMENDATION 1: The Department of the Air Force Chief Architect's Office and the Department of the Air Force Rapid Capabilities Office should define the Advanced Battle Management System (ABMS) architecture at the Joint All-Domain Command and Control level to ensure interoperability with other ABMS-like systems being developed. (Chapter 2)

RECOMMENDATION 2: The Joint Staff J6 or a designated U.S. Department of Defense executive agent should establish interoperability requirements and performance metrics for all participants in Joint All-Domain Command and Control to allow for eventual integration of all capabilities. (Chapter 2)

RECOMMENDATION 3: The Department of the Air Force Chief Architect's Office and the Department of the Air Force Rapid Capabilities Office should design the Advanced Battle Management System architecture to be modular and include open standards and interfaces that would enable configuration with other Service variants. (Chapter 2)

RECOMMENDATION 4: The Department of the Air Force Chief Architect's Office and the Department of the Air Force Rapid Capabilities Office should design the Advanced Battle Management System's architecture with specific technical requirements and solutions for ensuring that communications, data, and computation may continue to operate in degraded or denied access environments. (Chapter 2)

RECOMMENDATION 5: The Department of the Air Force Rapid Capabilities Office should adopt an array of data-exchange technologies that could support the entire spectrum of capabilities, from tactical to strategic. (Chapter 2)

RECOMMENDATION 6: To the maximum extent possible, the Department of the Air Force Chief Architect's Office and the Department of the Air Force Rapid Capabilities Office should design and execute a comprehensive artificial intelligence strategy that would encompass all elements, to include doctrine, chain of command, policy, authorization for weapon release in a joint environment, interfaces to Joint All-Domain Command and Control, and not just select capabilities of the Advanced Battle Management System. (Chapter 2)

RECOMMENDATION 7: The Joint All-Domain Command and Control cross functional team should reach immediate agreement on a common data fabric and security levels of the data with data standards and tools defined at the Joint level. Without a common set of agreed upon open standards with known interface exchange requirements that do not limit innovation, the military Services risk developing incompatible and stovepiped solutions. (Chapter 2)

RECOMMENDATION 8: In coordination with the Department of the Air Force Chief Software Officer, the Department of the Air Force Chief Architect's Office and the Department of the Air Force Rapid Capabilities Office should expand the use of containerization and Kubernetes for continuous Advanced Battle Management System development and for detecting and mitigating security vulnerabilities. (Chapter 2)

RECOMMENDATION 9: The Department of the Air Force Chief Architect's Office and the Department of the Air Force Rapid Capabilities Office should adopt development, security, and operations as the common development environment using containerization and continuous integration/continuous delivery across all of the Advanced Battle Management System. (Chapter 2)

RECOMMENDATION 10: For modular open-system designs with robust interface specifications, the Department of the Air Force Rapid Capabilities Office should acquire performance and interface requirements instead of all intellectual property rights. (Chapter 2)

RECOMMENDATION 11: The Department of the Air Force Chief Architect's Office and the Department of the Air Force Rapid Capabilities Office should design resilience into the Advanced Battle Management System architecture and specify dynamic criteria for needed performance. (Chapter 2)

RECOMMENDATION 12: The Joint Staff's J6, the Department of the Air Force, and the broader U.S. Department of Defense community should establish and implement a robust enterprise-wide offensive and defensive cybersecurity strategy for Joint All-Domain Command and Control (JADC2) and the Advanced Battle Management System. Security is a fundamental requirement that must be designed and fully integrated into the all JADC2-supporting systems' architecture from the start. (Chapter 2)

RECOMMENDATION 13: The Department of the Air Force Rapid Capabilities Office should apply zero trust (ZT) trust in stages as technologies mature and integrate ZT services to include the use of multi-factor authentication across all of the Advanced Battle Management System. (Chapter 2)

RECOMMENDATION 14: In addition to adopting zero trust, the Department of the Air Force Rapid Capabilities Office should leverage the best available mature cybersecurity practices and capabilities, including multi-factor authentication; identity, credential, and access management; encryption; penetration testing; managed detection services; behavior monitoring applications; among others. (Chapter 2)

RECOMMENDATION 15: The Department of the Air Force Rapid Capabilities Office (DAF RCO) should employ the Air Force's Mission Defense Teams to red team the Advanced Battle Management System's cyber defenses against attacks from malicious actors. Based on these red team exercises, the DAF RCO should address vulnerabilities by bolstering and enhancing cyber defenses accordingly. (Chapter 2)

RECOMMENDATION 16: The Department of the Air Force Chief Architect's Office and the Department of the Air Force Rapid Capabilities Office should work in partnership with the U.S. Cyber Command to address Internet of Things defense and other cyber vulnerabilities and exploits that are highlighted in the "United States Cyber Command Technical Challenge Problem Set" document. (Chapter 2)

RECOMMENDATION 17: The Department of the Air Force Chief Architect's Office and the Department of the Air Force Rapid Capabilities Office should work with the Department of the Air Force's Digital Engineering Enterprise Office to apply model-based systems engineering (MBSE) methods across Advanced Battle Management System engineering and sustainment activities and to enable MBSE to serve as a bridge between operator requirements and development teams. (Chapter 2)

RECOMMENDATION 18: Building on existing activities in digital engineering and modeling and simulations, the Department of the Air Force Chief Architect's Office and the Department of the Air Force Rapid Capabilities Office should expand the use of digital twins in Advanced Battle Management System development, particularly as new capabilities and technologies are introduced. (Chapter 2)

RECOMMENDATION 19: The Department of the Air Force Rapid Capabilities Office should consider scaling the Common Mission Control Center and designate it as phase zero for the Advanced Battle Management System. (Chapter 2)

Non-Technical

RECOMMENDATION 20: The Joint Staff J6 or a designated U.S. Department of Defense executive agent should establish an authoritative Joint-level body to address and resolve technical, operational, and command decisions for all contributors to the Joint All-Domain Command and Control framework. (Chapter 3)

RECOMMENDATION 21: The Joint Chiefs and military department secretaries should tackle the cultural, social, and emotional barriers to true Joint Warfighting Concept (JWC) horizontal integration if the Advanced Battle Management System and the larger Joint All-Domain Command and Control constructs are to enable the truly joint and multi-national integrated operations envisioned by the JWC. (Chapter 3)

RECOMMENDATION 22: The Department of the Air Force Rapid Capabilities Office should incorporate human systems integration methodologies into the Advanced Battle Management System to ensure that all human users are fully and effectively integrated with current and future systems elements.

RECOMMENDATION 23: The Department of the Air Force's Assistant Secretary of the Air Force for Acquisition, Technology, and Logistics and the Deputy Chief of Staff for Strategy, Integration, and Requirements should consider and weave personnel, cultural, training, and other non-materiel doctrine, organization, training, materiel, leadership, education, personnel, facilities, and policy issues into designs and implementation plans for the broader Advanced Battle Management System ecosystem. (Chapter 3)

RECOMMENDATION 24: Civilian and military leaders in the Department of the Air Force, Joint Staff, and the Office of the Secretary of Defense should ensure that the ethical use of artificial intelligence is examined and addressed in the Advanced Battle Management System's (and in other systems supporting the broader Joint All-Domain Command and Control framework's) design, operation, staffing, and training, as dictated by policy and the law of war. (Chapter 3)

RECOMMENDATION 25: The Air Education Training Command should establish a curriculum that would train or recruit highly qualified experts in artificial intelligence/machine learning, model-based systems engineering, cybersecurity, intelligence assessment, and test and evaluation for information technology, software, and hardware who can work with experts in military operations and culture. (Chapter 3)

CONCLUDING THOUGHTS

The vision and motivations for ABMS largely reflect conceptual and strategic needs in general terms. DoD often lacks the level of commonly available commercial information processing, storage, analysis, and sharing capabilities. ABMS on-ramp experimentations have demonstrated that such capabilities could be readily acquired to increase DAF and larger DoD capabilities.

However, the actual shortfalls of current systems in real terms, the potential gains from specific investments, and their operational implications are often classified, not clearly articulated, or yet to be determined. The ABMS effort would benefit greatly by making these more explicit—clearly articulating the *as-is* system (with its level of communication bandwidth, interconnectivities [or lack thereof], organizational and social interoperations, and the operational implications of the shortfalls) and the specific proposed investment options with their costs and operational benefits when proposing the next increment of the *to-be* system.[6] These are the type of specific investment options being considered by the DAF RCO in their capability releases, but a broad sense of this option space needs to be developed and articulated to all stakeholders, including Congress.

---

[6] Descriptions could include these example patterns. Weapon system X cannot currently talk to weapon system Y; the operational implication of this limitation is I, and here are the cost and benefits of solving this tactical communication shortfall. Operations Center A must manually correlate information from sensors Z and W to detect incoming threats, but this takes L times longer than it takes for the threat to complete its attack. The bandwidth of Operations Center O can handle only P percentage of available intelligence and sensor information, but capacity can be readily augmented through cloud services at cost C with the improved capacity to detect and process T more threats.

In addition to the technical and governance challenges laid out in this report, there are examples of specific operational gaps and shortfalls that reveal and reinforce the need to improve joint command, control, communications, computers, intelligence, surveillance, reconnaissance (C4ISR), and sensor-to-shooter capabilities. The committee is concerned, however, that the inability to articulate the magnitude of these operational shortfalls, their implications for national defense, the range of investment options (mature and developmental), and their costs and operational benefits may lead to insufficient attention and resources, as well as inadequate attention to the larger non-technical challenges that must be addressed if the Joint Warfighting Concept (JWC) is to be realized. The magnitude of the C4ISR shortfalls may not be reflected in the size or urgency of investments in ABMS and JADC2—in part because of a lack of clarity.

The committee also notes that DoD is often criticized for not being agile in acquiring and keeping pace with new technologies. The ABMS on-ramp experiments demonstrate the DAF's engagement in diverse, non-traditional commercial software and infrastructure companies through agile development and prototyping. Part of the challenge of fixing requirements, design, and budgets for ABMS is that such agility conflicts with more static approaches to acquisition. Nevertheless, some level of specificity is possible (often at the classified level). For example, what are the bandwidth and processing shortfalls at specific commands, what is the cost for acquiring secure cloud (or on-premises) capabilities to bring the commands up to modern levels, and what are the resulting increases in operational capabilities? Some specificity is needed for Congress and DoD to make trade-offs and understand exactly what ABMS can and should do.

ABMS and the larger JADC2 is a major undertaking—not only technically but also organizationally, socially, and ethically. Truly joint and seamless military operation has been a vision for many decades. New advances and insights have brought the DAF and the larger DoD to the verge of realizing this vision. However, it will take dedication, cooperation, grounded reality, planning, budgeting, and a willingness to seriously tackle the broader social and ethical aspects of such an endeavor. Experiments have shown the feasibility of some steps in this direction. The rest resides with leadership and teams to address these challenges.

Selected Bibliography

Allvin, D. 2021. "Why We Need the Advanced Battle Management System." DefenseOne. https://www.defenseone.com/ideas/2021/05/why-we-need-advanced-battle-management-system/173861/. May 6.

Brown, C.Q., Jr. 2020. *Accelerate Change or Lose*. https://www.af.mil/Portals/1/documents/2020SAF/ACOL_booklet_FINAL_13_Nov_1006_WEB.pdf. August.

CRS (Congressional Research Service). 2020. *The Army's Project Convergence*. https://sgp.fas.org/crs/weapons/IF11654.pdf. October 8.

CRS. 2020. "Defense Primer: Ground Based Strategic Deterrent (GBSD) Capabilities." Congressional Research Service in Focus. https://sgp.fas.org/crs/natsec/IF11681.pdf. November 10.

CRS. 2021. *Advanced Battle Management System (ABMS)*. June 29. https://sgp.fas.org/crs/weapons/IF11866.pdf. September 27.

CRS. 2021. *Joint All-Domain Command and Control (JADC2)*. https://sgp.fas.org/crs/natsec/IF11493.pdf. July 1.

CRS. 2021. *Joint All-Domain Command and Control: Background and Issues for Congress*. https://crsreports.congress.gov/product/pdf/R/R46725/2. March 18.

DAF (U.S. Department of the Air Force). 2020. *Department of the Air Force Manual 13-1AOC, Volume 3, Nuclear Space, Missile Command and Control Operational Procedures—Air Operations Center (AOC) Operations Center (OC)*. https://static.e-publishing.af.mil/production/1/af_a3/publication/dafman13-1aocv3/dafman13-1aocv3.pdf. December 18.

DDR&E (Department of Defense Research and Engineering Enterprise). 2021. "Human Systems Integration." https://ac.cto.mil/hsi/. September 18.

DoD (U.S. Department of Defense). 2018. *Summary of the 2018 National Defense Strategy of the United States of America: Sharpening the American Military's Competitive Edge*. https://dod.defense.gov/Portals/1/Documents/pubs/2018-National-Defense-Strategy-Summary.pdf.

DoD. 2019. *DoD Digital Modernization Strategy*. https://media.defense.gov/2019/Jul/12/2002156622/-1/-1/1/DOD-DIGITAL-MODERNIZATION-STRATEGY-2019.PDF. July 12.

DoD. 2020. *DoD Data Strategy*. https://media.defense.gov/2020/Oct/08/2002514180/-1/-1/0/DOD-DATA-STRATEGY.PDF.

GAO (Government Accountability Office). 2020. Defense Acquisitions: Action Is Needed to Provide Clarity and Mitigate Risks of the Air Force's Planned Advanced Battle Management System. https://www.gao.gov/assets/gao-20-389.pdf. April.

Selected Bibliography

NSCAI (National Security Commission on Artificial Intelligence). 2021. *National Security Commission on Artificial Intelligence Final Report*. https://www.nscai.gov/wp-content/uploads/2021/03/Full-Report-Digital-1.pdf. March.

Office of the Under Secretary of Defense (Comptroller/Chief Financial Officer). 2021. *Defense Budget Overview, United States Department of Defense Fiscal Year 2022 Budget Request*. https://comptroller.defense.gov/Portals/45/Documents/defbudget/FY2022/FY2022_Budget_Request_Overview_Book.pdf. May.

Roper, W. 2020. "Advanced Battle Management System Management Construct." Memorandum for Record, Office of the Assistant Secretary of the Air Force for Acquisition, Technology, and Logistics. https://insidedefense.com/sites/insidedefense.com/files/documents/2020/nov/11242020_abms.pdf. November 24.

Roper, W. 2020. "There Is No Spoon: The New Digital Acquisition Strategy." https://software.af.mil/wp-content/uploads/2020/10/There-Is-No-Spoon-Digital-Acquisition-7-Oct-2020-digital-version.pdf. October 7.

Rose, S., O. Borchert, S. Mitchell, and S. Connelly. 2020. "Zero Trust Architecture." NIST Special Publication 800-207. https://nvlpubs.nist.gov/nistpubs/SpecialPublications/NIST.SP.800-207.pdf. August.

Snyder, D., J.D. Powers, E. Bodine-Baron, B. Fox, L. Kendrick, and M.H. Powell. 2015. "Improving the Cybersecurity of U.S. Air Force Military Systems Throughout Their Life Cycles." RAND Corporation. https://www.rand.org/content/dam/rand/pubs/research_reports/RR1000/RR1007/RAND_RR1007.pdf.

Appendixes

A

Statement of Task

The National Academies of Sciences, Engineering, and Medicine will establish a committee to plan and conduct a classified study to assess the planned Advanced Battle Management System (ABMS) communications and systems integration architecture. The study will examine the technical approach being employed by ABMS and its ability to effectively support the range of system integration desired, while also supporting operational and development agility; and the governance being applied by ABMS and if it is appropriate and sufficient to enable quick development and evolution of capabilities while maintaining appropriate government control over the output. Specifically, the committee will:

1. Evaluate the planned ABMS data and communication architecture and compare the architecture anticipated performance characteristics to those needed to support real-time fire control and all-domain sensor-to-shooter data flow, command and control (C2) activities, artificial intelligence (AI)-based patterns-of-life training, battle damage assessment, and other related data-based activities.
2. Determine any technical gaps and shortfalls in ABMS technology and planned system integration architecture.
3. Review ABMS governance and recommend how planned organizational and execution plans and processes may be improved to better enable a rapid realization of Joint All-Domain Command and Control (JADC2)

operations for the Department of the Air Force and the U.S. Department of Defense, as a whole.

The committee will convene a data-gathering workshop and four meetings of the study team and other attendees, as required, to gain relevant information. The committee will also provide a classified report summarizing the results from the study, with an unclassified public summary.

B

Data-Gathering Meetings

COMMITTEE MEETING 1
OCTOBER 30, 2020

1130–1300 Perspectives from Study Co-Sponsors

           Mr. Preston Dunlap, Chief Architect, U.S. Department of the Air Force
           Ms. Ally Bain, Program Examiner, National Security Division, Office of Management and Budget

COMMITTEE MEETING 2
DECEMBER 18, 2020

1430–1440 Welcome Remarks

           Dr. Philip Antón, Committee Chair

1440–1600 Joint Warfighting Concept: Joint All-Domain Command and Control and the Advanced Battle Management System

           Brig Gen Jeffery D. Valenzia, Director, Joint Force Integration, Deputy Chief of Staff for Strategy, Integration and Requirements, Headquarters U.S. Air Force

COMMITTEE MEETING 3
JANUARY 8, 2021

1430–1440 Welcome Remarks

Dr. Philip Antón, Committee Chair

1440–1600 U.S. Army's Project Convergence

Col Andre' (Dre') B. Abadie, Ph.D., Solutions Architect, Networks-AI-Cyber and Project Convergence, Army G3/5/7, U.S. Army Futures Command

COMMITTEE MEETING 4
JANUARY 22, 2021

1430–1440 Welcome Remarks

Dr. Philip Antón, Committee Chair

1440–1600 Perspectives from the Air Force Rapid Capabilities Office

Mr. Randy Walden, Director and Program Executive Officer, Air Force Rapid Capabilities Office, Office of the Assistant Secretary of the Air Force for Acquisition, Technology and Logistics, Washington, DC

COMMITTEE MEETING 5
FEBRUARY 5, 2021

1430–1445 Welcome Remarks

Dr. Philip Antón, Committee Chair

1445–1600 Joint All-Domain Command and Control (JADC2)

Mr. Stuart A. Whitehead, Deputy Director, Cyber and C4 Integration, Joint Staff J6

Appendix B

Mr. John S. Wellman, Deputy Director, Command and Control Integration, Chief, Command and Control Capabilities Division, Joint Staff J6

COMMITTEE MEETING 6
FEBRUARY 24, 2021

1100–1115	Welcome Remarks

Dr. Philip Antón, Committee Chair

1115–1230	DoD C3I Perspectives

Hon. John P. Stenbit, Director, Viasat, and Former Assistant Secretary of Defense for Command, Control, Communications, and Intelligence

1230–1345	Joint Common Foundation and JAIC's Decision Engineering Process

Mr. Nand Mulchandani, Chief Technology Officer, Joint Artificial Intelligence Center (JAIC), U.S. Department of Defense

1345–1400	Break
1400–1500	U.S. Northern Command Support to ABMS

Col Matt "Nomad" Strohmeyer, USAF, NORAD/USNORTHCOM J8 JADC2 Development Lead

1500–1600	SolarWinds Cyber Breach

Mr. Matthew Butkovic, Technical Director, Cyber Risk and Resilience Assurance
Dr. Robert Cunningham, Associate Director, Cyber Assurance
Mr. Art Manion, Technical Manager, Vulnerability Analysis, CERT Division, Software Engineering Institute, Carnegie Mellon University

1600–1700	Air Force Enterprise IT to Enable ABMS
	Ms. Lauren Knausenberger, Chief Information Officer, Department of the Air Force

COMMITTEE MEETING 7
MARCH 3, 2021

1100–1115	Welcome Remarks
	Dr. Philip Antón, Committee Chair
1115–1245	Department of the Navy All-Domain Operations
	Hon. James F. Geurts, Performing the Duties of the Under Secretary of the Navy
1245–1300	Break
1300–1415	Joint All-Domain Command and Control
	Lt Gen Dennis A. Crall, USMC, Director, Command, Control, Communications and Computers/Cyber, and Chief Information Officer, Joint Staff, J6
1415–1430	Break
1430–1600	Naval Digital Strategy
	Ms. Kelly McCool, Director, Digital Warfare Office, OPNAV N9 and N2N6
1600–1700	Project Overmatch
	RADM Douglas W. Small, USN, Commander, Naval Information Warfare Systems Command

Appendix B

COMMITTEE MEETING 8
MARCH 5, 2021

1430–1445 Welcome Remarks

Dr. Philip Antón, Committee Chair

1445–1600 Software Pathways

Dr. Forrest Shull, Lead for Defense Software Acquisition Policy Research, Carnegie Mellon University Software Engineering Institute

COMMITTEE MEETING 9
MARCH 19, 2021

1430–1440 Welcome Remarks

Dr. Philip Antón, Committee Chair

1440–1600 ACC Perspectives on ABMS

Mr. John F. Vona, Deputy Director of the Plans, Program and Requirements Directorate (ACC/A5/8/9), Headquarters Air Combat Command

Dr. John D. Matyjas, Scientific Advisor to the Commander, Headquarters Air Combat Command

Col Walter C. Hattemer, Chief, Command and Control Weapons Systems Division, Headquarters Air Combat Command

Col (Ret.) Dennis P. (Devo) McDevitt, Deputy Chief, C2ISR Operations Division, Headquarters Air Combat Command

Lt Col George M. Hart III, Deputy Director, ISR Mission and Capabilities Division, Intelligence Directorate, Headquarters Air Combat Command

Lt Col Keith C. McGuire, Chief, Airborne Command and Control Weapons System Requirements Branch, Headquarters Air Combat Command

COMMITTEE MEETING 10
MARCH 30–31, 2021

0900–1630   Classified Data-Gathering Session

COMMITTEE MEETING 11
APRIL 16, 2021

1430–1440   Welcome Remarks

                Dr. Philip Antón, Committee Chair

1440–1600   Cybersecurity in JADC2 and Contested Environments

                Mr. Eric Bryant (DISL), Technical Director, Weapons and Space Cybersecurity, National Security Agency

COMMITTEE MEETING 12
APRIL 21, 2021

1000–1015   Welcome Remarks

                Dr. Philip Antón, Committee Chair

1015–1200   FFRDC and UARC Panel on ABMS

                Dr. Mark Happel, Supervisor, Data Science and Machine Learning Section, Johns Hopkins University Applied Physics Laboratory
                Dr. Sherrill Lingel, Senior Engineer, RAND Corporation
                Dr. Jennifer Watson, Assistant Division Head for ISR, MIT Lincoln Laboratory

1200–1330   Testing and Evaluation in ABMS

                Maj Gen Christopher P. Azzano, Commander, Air Force Test Center, Edwards Air Force Base

Appendix B

1330–1430	Industry Perspectives: Chooch AI's Support to ABMS On-Ramp 2
	Mr. Drew Fanning, Vice President, Chooch AI
1430–1600	Industry Perspectives: Microsoft
	Mr. Scott Stebbins, Digital Advisor/Architect, Defense, Microsoft Corporation
	Mr. Derek Strausbaugh, Chief Digital Officer, Defense, Microsoft Corporation
	Mr. John Vargas, Air and Space Force Account Executive, Microsoft Corporation

COMMITTEE MEETING 13
MAY 14, 2021

1430–1435	Welcome Remarks
	Dr. Philip Antón, Committee Chair
1435–1600	FFRDC Perspectives on ABMS
	Mr. Scott Lee, Cross-Cutting Priority Lead for Joint All-Domain Command and Control (JADC2), MITRE Corporation
	Maj Gen (Ret.) Martin Whelan, USAF, Senior Vice President of Defense Systems Group, The Aerospace Corporation

C

Acronyms and Abbreviations

A2/AD	anti-access/area denial
AADC	Area Air Defense Commander
ABMS	Advanced Battle Management System
ACA	airspace control authority
ACC	Air Combat Command
ADSV	ABMS DeviceONE SecureView
AFC	Army Futures Command
AFMC	Air Force Materiel Command
AFRL	Air Force Research Laboratory
AFSC	Air Force Specialty Code
AI	artificial intelligence
AOC	Air Operations Center
AOC-WS	Air Operations Center—Weapon System
API	Application Program Interface
ARB	architecture review board
ATO	Air Tasking Order
AutoML	automated machine learning
AWACS	Airborne Warning and Control System
BPM	business process management

Appendix C

C2	command and control
C3I	command, control, communications, and intelligence
C4ISR	command, control, communications, computers, intelligence, surveillance, reconnaissance
CACE	changing anything changes everything
CAO	Chief Architect's Office
CCSDS	Consultative Committee for Space Data Systems
CFT	cross functional team
CI/CD	continuous integration/continuous delivery
CIO	Chief Information Officer
CMCC	Common Mission Control Center
COP	common operating picture
CR1	capability release one
CR2	capability release two
CSO	Chief Software Officer
CSTE	common simulation training environment
DAF	Department of the Air Force
DAF RCO	Department of the Air Force Rapid Capabilities Office
DARPA	Defense Advanced Research Projects Agency
DAU	Defense Acquisition University
DDR&E	U.S. Department of Defense Research and Engineering Enterprise
DE	digital engineering
DevSecOps	development, security, and operations
DFARS	Defense Federal Acquisition Regulation Supplement
DISA	Defense Information Systems Agency
DoD	U.S. Department of Defense
DoN	Department of the Navy
DOTMLPF-P	doctrine, organization, training, materiel, leadership, education, personnel, facilities, and policy
DSOP	U.S. Department of Defense Enterprise DevSecOps Initiative
DTO	Digital Transformation Office
EW	electronic warfare
FFTTEA	find, fix, target, track, engage, and assess
FNC3	fully networked command, control, and communications
FOS	family of systems
FY	fiscal year

GAO	Government Accountability Office
GBSD	ground-based strategic deterrent
GPS	global positioning system
HCI	human-computer interface
HFE	human factors engineering
HQ	headquarters
HQE	highly qualified expert
HSI	human system integration
HUD	heads-up display
iBPMS	intelligent business process management suites
ICAM	identity, credential, and access management
IEEE	Institute of Electrical and Electronics Engineers
IO	information operations
IoT	Internet of Things
IP	intellectual property
iPaaS	integration platform as a service
ISR	intelligence, surveillance, reconnaissance
IT	information technology
JADC2	Joint All-Domain Command and Control
JAPCC	Joint Air Power Competence Centre
JFACC	Joint Forces Air Component Commander
JFC	Joint Force Commander
JROC	Joint Requirements Oversight Council
JSTARS	Joint Surveillance and Target Attack Radar System
JWC	Joint Warfighting Concept
KREL	Kessel Run Experimentation Lab
LOE	line of effort
M&S	modeling and simulation
MBSE	model-based systems engineering
MFA	multi-factor authentication
ML	machine learning
MLS	multi-level security
MOSA	modular open systems architecture

Appendix C

NATO	North Atlantic Treaty Organization
NC3	nuclear command, control, communications
NDS	National Defense Strategy
NGAD	next generation air dominance
NORAD	North American Aerospace Defense Command
NSCAI	National Security Commission on Artificial Intelligence
OMG	Object Management Group
OODA	observe-orient-decide-act
OPR	office of primary responsibility
OSA	open system architecture
OSD	Office of the Secretary of Defense
OUSD A&S	Office of the Under Secretary of Defense for Acquisition and Sustainment
PATS	protected anti-jam tactical satellite communications
PC	Project Convergence
PEO	program executive office
PRC	People's Republic of China
R&D	research and development
RDA	research, development, acquisition
RPA	robotic process automation
SAE	Service Acquisition Executive
SAF/AQ	Assistant Secretary of the Air Force for Acquisition, Technology, and Logistics
SATCOM	satellite communication
SCARS	simulator common architectures requirements and standards
SEP	systems engineering process
shOC-N	Shadow Operations Center-Nellis
SIPRNET	secret Internet protocol router network
SOCOM	Special Operations Command
SOF	Special Operations Forces
SOFIC	Special Operations Forces industry conference
SOH	safety and occupational health
STITCHES	system of systems technology integration tool chain for heterogeneous electronic systems

T&E	testing and evaluation
TMD	theater missile defense
TTA	time-triggered architecture
TTP	tactics, techniques, and procedures
UCI	universal command and control interface
UDL	Unified Data Library
USCYBERCOM	U.S. Cyber Command
USN	U.S. Navy
USNORTHCOM	U.S. Northern Command
USSF	U.S. Space Force
USSPACECOM	U.S. Space Command
USSTRATCOM	U.S. Strategic Command
VV&A	verification, validation, and accreditation
W1	WeaponONE
ZT	zero trust

D

Committee Member Biographical Information

PHILIP S. ANTÓN, *Chair*, is the chief scientist at the Acquisition Innovation Research Center at the Stevens Institute of Technology. Previously, Dr. Antón was a senior information scientist at the RAND Corporation, where he conducted research on acquisition and sustainment policy; cybersecurity; emerging technologies; technology foresight; process performance measurement and efficiency; data science and analytics; aeronautics test infrastructure; and military modeling and simulation. From 2011 to 2016, Dr. Antón served two tours in the Pentagon, filling a senior executive service position directing the Acquisition Policy Analysis Center. Reporting directly to the Under Secretary of Defense for Acquisition Technology and Logistics, he conducted strategic initiatives to improve the performance of the U.S. Department of Defense's policies and institutions, crafted affordability policy, and brought new analytic insights into the performance of acquisition and sustainment policies, processes, and tradecraft. For these contributions, Dr. Antón received the Secretary of Defense Medal for Outstanding Public Service in 2017. From 2004 to 2011, Dr. Antón was the director of the Acquisition and Technology Policy Center in RAND's National Security Research Division. This center addressed how accelerating technological change and modernization efforts will transform the U.S. national security establishment. It also explored new acquisition and management strategies and ways to maintain core defense technology and production bases. Dr. Antón earned his M.S. and Ph.D. in information and computer science from the University of California, Irvine, specializing in computational neuroscience and artificial intelligence. His B.S. is in engineering from the University of California, Los Angeles, specializing in computer engineering.

SHARON A. BEERMANN-CURTIN is an independent consultant with more than 30 years of government experience in technology and product development. Prior to leaving government service, Ms. Beermann-Curtin served as the acting director and the deputy director in the Office of the Under Secretary of Defense—Research and Engineering, Strategic Capabilities Office (SCO), whose mission is to identify, analyze, and accelerate the development and transition of capabilities to counter strategic adversaries. In these roles, Ms. Beermann-Curtin grew the organization from a start-up task force to an office of innovation within the U.S. Department of Defense (DoD). Prior to joining SCO, she served as the technical lead for power and energy at the Office of Naval Research (ONR) between 2010 and 2014, managing the organization's high-power electrical ship systems, power source and conversion technologies, alternative fuels, and Future Naval Capabilities Power and Energy Pillar. In 2004, Ms. Beermann-Curtin joined the Defense Advanced Research Projects Agency, serving as a program manager for 5 years. She served in both the Defense Sciences Office and the Microsystems Technology Office, with a portfolio of programs focused on power and energy generation and electrical system components, including batteries, fuel cells, high-power capacitors, high-power semiconductors (silicon carbide based), and biofuels through chemical synthesis (sunlight to fuel). From 2002 to 2003, she was a visiting scholar to the Massachusetts Institute of Technology. Ms. Beermann-Curtin has a vast knowledge of DoD acquisition, serving from 1999 to 2001 as the first chief technology officer for the Program Executive Office—Aircraft Carriers, responsible for the transition of new technologies to both in-service and future aircraft carriers. She also held numerous positions at ONR, including Acting Deputy Department Head of the Materials and Physicals Sciences and Ship Hull Mechanical and Electrical Science and Technology (S&T) Department; Technology Manager for Ship Systems in the Hull, Mechanical and Electrical S&T Division; and Program Manager for Underwater Weapons Countermeasures. Ms. Beermann-Curtin holds an M.S. in electrical engineering from the University of Rhode Island and a B.S. in electrical engineering from Missouri University of Science and Technology.

MICHAEL A. FANTINI retired from the U.S. Air Force as a Major General after a 34-year career. Gen. Fantini most recently served as the acting deputy chief of staff for strategy, integration, and requirements and the director of Air Force Warfighting Integration Capability (AFWIC), where he led enterprise-wide integration and future force design to enable the Air Force to rapidly transition into a networked, multi-domain 21st century force. Prior to his AFWIC assignment, he was the director of Global Power Programs in the Office of the Secretary of the Air Force for acquisition, technology, and logistics (SAF/AQ). Previously, Gen. Fantini served as Commander, Kandahar Airfield (COMKAF-NATO), Afghanistan. As COMKAF, he was responsible for the operational efficiency and readiness of Kandahar Airfield. He acted as senior airfield authority and a task force commander-equivalent in defense of the airfield, exercising centralized coordination of airfield operations, logistics, NATO assets, and real estate management, while leading all force protection actions in defense of nearly 22,000 assigned and attached personnel. Gen. Fantini has served in a variety of operational assignments as an F-16 pilot, instructor pilot, and weapons officer. He has commanded a fighter squadron, the 332nd Expeditionary Operations Group in Balad, Iraq; the 82nd Training Wing, Sheppard Air Force Base, Texas; and the 451st Air Expeditionary Wing, Kandahar, Afghanistan. He has served in multiple staff positions, including Chief of the Fighter Weapons Branch, Secretary of the Air Force Office of Special Programs; Operations Officer and Deputy Division Chief of Global Force Management at the Joint Operations Directorate; and Director, Combat Force Application and Operational Capabilities Requirements. Gen. Fantini earned his B.S. in mechanical engineering from the Catholic University of America, a master's degree in aviation science from Embry-Riddle University, and a master's degree in national security studies from the National War College. Gen. Fantini was a command pilot with more than 3,400 hours in the MQ-9, F-16, T-37, and T-38.

PRISCILLA E. GUTHRIE is a fellow in the Information Technology and Systems Division at the Institute for Defense Analyses. Previously, Ms. Guthrie served as the special command advisor, Cyber Security/Information Technology/Information Assurance for U.S. Cyber Command. Prior to that, she was a vice president at ViaSat, Inc. In 2009, Ms. Guthrie was confirmed by the Senate as the chief information officer in the Office of the Director of National Intelligence (ODNI). She also served as the director of the Information Technology and Systems Division at IDA and the Deputy Assistant Secretary of Defense (Deputy Chief Information Officer) at the U.S. Department of Defense (DoD). Before moving to the Pentagon, Ms. Guthrie was a vice president of TRW, Inc., where she led business units in defense, intelligence, automotive, and information technology. Ms. Guthrie supports various advisory groups for DoD, primarily in the areas of cybersecurity and information technology, including the U.S. Strategic Command's Strategic Advisory Group, the Defense Science Board, and several outside advisory boards, including Penn State's Outreach and Online Advisory Board and the Society of Distinguished Alumni executive board. She has an M.B.A. from Marymount College and a B.S.E.E. from Pennsylvania State University.

PAUL G. KAMINSKI, NAE, is the chair and the chief executive officer of Technovation, Inc., a consulting company dedicated to fostering innovation and the development and application of advanced technology. Dr. Kaminski is a former Under Secretary of Defense (Acquisition, Technology, and Logistics) and was responsible for all U.S. Department of Defense (DoD) research, development, and acquisition programs. During his Air Force career, he served as director for low observable technology, with responsibility for overseeing the development, production, and fielding of major "stealth" systems (e.g., F-117, B-2). He also led the initial development of a National Reconnaissance Office space system and related sensor technology. Dr. Kaminski's government advisory memberships have included the Senate Select Committee on Intelligence Technical Advisory Board, the Defense Science Board (chairman two times) the President's Intelligence Advisory Board, the Director for National Intelligence's Senior Advisory Group, and the FBI Director's Advisory Board. He is a fellow of the Institute of Electrical and Electronics Engineers (IEEE) and a fellow and honorary fellow of the American Institute of Aeronautics and Astronautics. Dr. Kaminski has authored numerous publications dealing with inertial and terminal guidance system performance, simulation techniques, and Kalman filtering and numerical techniques applied to estimation problems. He received a B.S. from the Air Force Academy, an M.S. in both aeronautics and astronautics and in electrical engineering from the Massachusetts Institute of Technology, and a Ph.D. in aeronautics and astronautics from Stanford University. Dr. Kaminski received the National Medal of Technology in 2006, U.S. Department of Defense Medal for Distinguished Public Service—5 awards, Defense Distinguished Service Medal, Director of Central Intelligence Director's Award, DIA Director's Award, Air Force Academy 2002 Distinguished Graduate Award, the Ronald Reagan Award for Missile Defense, the Perry Award for precision strike, the Reed Award for Aeronautics, the IEEE Simon Ramo Award for Systems Engineering, and the IISS Possony Medal for Outstanding Contributions to Strategic Progress through Science and Technology. He is a member of the National Academy of Engineering and was elected to the National Aviation Hall of Fame in 2020.

THOMAS A. LONGSTAFF is the chief technology officer (CTO) at Carnegie Mellon University's Software Engineering Institute (SEI). As CTO, Dr. Longstaff is responsible for formulating a technical strategy and leading the funded research program of the institute based on current and predicted future trends in technology, government, and industry. Before joining the SEI as CTO in 2018, Dr. Longstaff was a program manager and the principal cybersecurity strategist for the Asymmetric Operations Sector of the Johns Hopkins University Applied Physics Laboratory (APL), where he led projects on behalf of the U.S. government, including nuclear command and control, automated incident response, technology transition of cyber research and development, information assurance, intelligence, and global information networks. He also was the chair of the Computer Science, Cybersecurity, and Information Systems Engineering Programs and the co-chair of Data Science in the Whiting School at Johns Hopkins. Dr. Longstaff's academic publications span topics such as malware analysis, information survivability, insider threat, intruder modeling, and intrusion detection. He maintains an active role in the information assurance community and regularly advises organizations on the future of network threat and information assurance. He is an editor for *Computers and Security*, and has previously served as associate editor for *IEEE Security and Privacy*; general chair for the New Security Paradigms Workshop and Homeland Security Technology Conference; and numerous other program and advisory committees. Prior to joining the staff at APL, Dr. Longstaff was the deputy director for technology for the Computer Emergency Response Team (CERT) Division at the Software Engineering Institute. In his 15-year tenure at the SEI CERT Division, he helped create many of the projects and centers that made the program an internationally recognized network security organization. His work included assisting the U.S. Department of Homeland Security and other agencies to use response and vulnerability data to define and direct a research and operations program in analysis and prediction of network security and cyber terrorism events. Dr. Longstaff received his bachelor's degree in physics and mathematics from Boston University and his master's degree in applied science and his Ph.D. in computer science from the University of California, Davis.

KATHARINA G. McFARLAND is a commissioner on the National Security Commission on Artificial Intelligence (NSCAI) and the chair of the Board on Army Research and Development at the National Academies of Sciences, Engineering, and Medicine. Mrs. McFarland retired in January 2017 as the Assistant Secretary of Defense for Acquisition and Acting Assistant Secretary of the Army (Acquisition, Logistics, and Technology) following designation by President Barack Obama on February 1, 2016. As the Assistant Secretary of Defense for Acquisition, confirmed in May 2012, she served as principle acquisition advisor to the Secretary of Defense and the Under Secretary of Defense for Acquisition, Technology, and Logistics on all U.S. Department of Defense and IC acquisition matters and had oversight of the Defense Acquisition University, the Human Capitol Office (All Acquisition Workforce), Program Assessment and Root Cause Assessment, and Defense Contract Management Agency. As the Assistant Secretary of the Army (Acquisition, Logistics, and Technology) and Army Acquisition Executive, Mrs. McFarland oversaw the execution of the Army's acquisition function, including life cycle management and sustainment of Army weapons systems and research and development programs, and managed the Army Acquisition Corps and greater Army Acquisition Workforce. Mrs. McFarland also served as the science advisor to the Secretary of the Army and as the Army's senior research and development official and senior procurement executive. Prior to these roles, she served as the president of the Defense Acquisition University (DAU). Under her leadership, DAU provided practitioner training, career management, and services to enable the acquisition, technology, logistics, and requirements community to make smart business decisions and deliver timely and affordable capabilities to the warfighter. Before joining DAU, Mrs. McFarland was the Director for Acquisition for the Missile Defense Agency (MDA), a position she held since May 2006. As MDA's principal acquisition executive, Mrs. McFarland advised the director of MDA on all acquisition, contracting, and small business decisions. Other core responsibilities included the development of process activities and program policy associated with the execution of the single integrated Ballistic Missile Defense System (BMDS) research, development, and test program, and establishment of the Baseline Execution Review to ensure that an integrated program execution of the BMDS occurred across the baselines of schedule, cost, performance, contracting, test and operational delivery. Mrs. McFarland began her civil service career in 1986 as a general engineer at Headquarters Marine Corps, where she was accredited as a materials, mechanical, civil, and electronics engineer. She has received an Honorary Doctoral of Engineering from the University of Cranfield, United Kingdom; the Presidential Meritorious Executive Rank Award; the Secretary of Defense Medal for Meritorious Civilian Service Award; the Department of the Navy Civilian Tester of the Year Award; and the Navy and U.S. Marine Corps Commendation Medal for Meritorious Civilian Service. Mrs. McFarland is Defense Acquisition Workforce Improvement Act Level-III-certified in program management, engineering, and testing as well as having a professional engineer license and having attained her Project Management Professional certification.

GUNASEKARAN SEETHARAMAN is the U.S. Navy senior scientist for advanced computing concepts and the chief scientist of computation, Center for Computational Sciences at the Naval Research Laboratory (NRL). Dr. Seetharaman has worked on computer vision, parallel computing, and machine perception algorithms for more than 30 years. He started his academic career at the University of Louisiana, Lafayette, in 1988. Dr. Seetharaman joined the Air Force Institute of Technology (AFIT) in 2003, and moved to the Air Force Research Laboratory in 2008 and to NRL in 2015. He also held visiting professor positions as a Centre National de la Recherche Scientifique (CNRS) research professor at the University of Paris XI and as a distinguished professor at the Indian Institute of Technology, Mumbai. At AFIT, Dr. Seetharaman worked in a team for prototyping a wide area motion imaging (WAMI) platform that was transitioned to theater. He collaboratively led Team Cajun-Bot and fielded two unmanned vehicles at the Defense Advanced Research Projects Agency grand challenge named Cajun-Bot and Rajin-Bot. He is a fellow of the Institute of Electrical and Electronics Engineers (IEEE), recognized for his contributions to high-performance computer vision algorithms for airborne imagery, and served as the elected section chair of IEEE Mohawk Valley Section Region 1. He was inducted into the Electronic Warfare Technology Hall of Fame in 2020, and is a member of Tau Beta Pi, Eta Kappa Nu, Upsilon Pi Epsilon, and Phi Beta Delta honor societies. Dr. Seetharaman serves as an associate editor of the *Association for Computing Machinery Journal of Computing Surveys*. He earned his Ph.D. in electrical and computer engineering from the University of Miami, M.Tech. in electrical engineering from the Indian Institute of Technology, Madras, and B.E. in electronics and communication engineering from the University of Madras.

DAVID M. VAN BUREN is the chief executive officer of Crossroads Management, a business strategy and program development firm. He sits on multiple boards and consults for high technology, as well as internationally owned firms. Prior to Crossroads, Mr. Van Buren served as the L3 Technologies senior vice president for program development. His responsibilities included corporate business strategy, corporate international operations, classified program development and security infrastructure, and corporate quality and continuous program improvement. Before joining L3, Mr. Van Buren was the Air Force service acquisition executive from 2009 to 2012, where he was responsible for all Air Force research, development, and acquisition activities. He directed approximately $70 billion of annual investments that included major programs such as the B-21, F-35, KC-46A, and all space and weapons programs, as well as information technology, cyber, command and control, and intelligence, surveillance, and reconnaisance systems. He was also responsible for initiating the classified family of systems in 2010. Mr. Van Buren possesses more than 40 years of business experience in the Air Force, large defense corporations, and private equity-owned small and medium aerospace and commercial high-technology firms. These technology areas included hyperspectral imaging, laser communications, alternative power sources, avionics, high-speed processing, compound semiconductors, and satellite power systems. Previously, Mr. Van Buren was the vice president and the deputy program manager for the B-2 bomber at Northrop Corporation, and project manager on several classified airborne platforms, including the F-117A at Lockheed. Prior to his tenure at Lockheed, he served on active duty in fighter operations and program management in the Air Force for 9 years, including two operational tours in Southeast Asia. Mr. Van Buren earned his B.A. in science and mathematics from the University of Illinois and his M.A. in industrial management from Central Michigan University. He has also completed an executive engineering program from Stanford University.

E

Disclosure of Unavoidable Conflicts of Interest

The conflict-of-interest policy of the National Academies of Sciences, Engineering, and Medicine (https://www.nationalacademies.org/about/institutional-policies-and-procedures/conflict-of-interest-policies-and-procedures) prohibits the appointment of an individual to a committee like the one that authored this Consensus Study Report if the individual has a conflict of interest that is relevant to the task to be performed. An exception to this prohibition is permitted only if the National Academies determine that the conflict is unavoidable and the conflict is promptly and publicly disclosed.

When the committee that authored this report was established, a determination of whether there was a conflict of interest was made for each committee member given the individual's circumstances and the task being undertaken by the committee. A determination that an individual has a conflict of interest is not an assessment of that individual's actual behavior or character or ability to act objectively despite the conflicting interest.

Dr. Paul Kaminski has a conflict of interest in relation to his service on the Committee on Air Force Advanced Battle Management System because of his current affiliation with or financial interests in defense sector companies that may pursue or have contracts in support of the Advanced Battle Management System (ABMS), including General Dynamics, The Boeing Company, and Raytheon.

Hon. Katharina McFarland has a conflict of interest in relation to her service on the Committee on Air Force Advanced Battle Management System because of her current service on the Board of Directors of the Science Applications International Corporation and Raytheon, which may have or pursue contracts in support of the ABMS.

In each case, the National Academies determined that the experience and expertise of the individuals were needed for the committee to accomplish the task for which it was established. The National Academies could not find other available individuals with the equivalent experience and expertise who did not have a conflict of interest. Therefore, the National Academies concluded that the conflict was unavoidable and publicly disclosed it on its website (https://www.nationalacademies.org/our-work/advanced-battle-management-system-consensus-study).